NEW RESEARCH ON BIOFUELS

NEW RESEARCH ON BIOFUELS

JAMES H. WRIGHT AND DANIEL A. EVANS
EDITORS

Nova Science Publishers, Inc.
New York

Copyright © 2008 by Nova Science Publishers, Inc.

For permission to use material from this book please contact us:
Telephone 631-231-7269; Fax 631-231-8175
Web Site: http://www.novapublishers.com

NOTICE TO THE READER

The Publisher has taken reasonable care in the preparation of this book, but makes no expressed or implied warranty of any kind and assumes no responsibility for any errors or omissions. No liability is assumed for incidental or consequential damages in connection with or arising out of information contained in this book. The Publisher shall not be liable for any special, consequential, or exemplary damages resulting, in whole or in part, from the readers' use of, or reliance upon, this material. Any parts of this book based on government reports are so indicated and copyright is claimed for those parts to the extent applicable to compilations of such works.

Independent verification should be sought for any data, advice or recommendations contained in this book. In addition, no responsibility is assumed by the publisher for any injury and/or damage to persons or property arising from any methods, products, instructions, ideas or otherwise contained in this publication.

This publication is designed to provide accurate and authoritative information with regard to the subject matter covered herein. It is sold with the clear understanding that the Publisher is not engaged in rendering legal or any other professional services. If legal or any other expert assistance is required, the services of a competent person should be sought. FROM A DECLARATION OF PARTICIPANTS JOINTLY ADOPTED BY A COMMITTEE OF THE AMERICAN BAR ASSOCIATION AND A COMMITTEE OF PUBLISHERS.

LIBRARY OF CONGRESS CATALOGING-IN-PUBLICATION DATA

New research on biofuels / James H. Wright and Daniel A. Evans (editors).
   p. cm.
 ISBN 978-1-60456-828-8 (hardcover)
 1. Biomass energy. I. Wright, James H., 1964- II. Evans, Daniel A.
 TP339.N49 2008
 662'.88--dc22
         2008027620

Published by Nova Science Publishers, Inc. ✦ New York

CONTENTS

PREFACE

This book provides new research on biofuels from around the globe.

Chapter 1 - The production of ethanol by yeast is a key technology in ethanol fermentation. However, ethanol is toxic to yeast cells. It causes misfolding of proteins, an increase in membrane fluidity, changes in mRNA export from the nucleus, and activation of various stress signaling pathways, including the protein kinase A pathway. Ultimately, it causes cell death. Ethanol-induced death in yeast occurs during the fermentation of several alcoholic beverages such as sake (Japanese rice wine) and wine. In contrast, during most of the fermentation of bioethanol yeast cells do not die, either because, the ethanol concentration does not reach a high enough level, or because the fermentation period is not long enough to cause the yeast to die. However, future bioethanol production technology may require more severe fermentation conditions. Such technology could include the production of ethanol to a higher ethanol concentration, repetitive fermentation by fixed yeast cells, or fermentation at high temperature. Thus, ethanol-induced death during fermentation may also be a limiting factor in the future production of bioethanol. Therefore, elucidation of the mechanism of ethanol-induced death merits further research. This knowledge could be used for the development of technology to enhance yeast productivity in the fermentation of alcoholic beverages and bioethanol.

Cell death can be grouped into two categories—necrosis and apoptosis. While necrosis is a catastrophic and passive death, apoptosis is a programmed death that is of benefit to the organism as a whole. Historically, apoptosis was first identified as an altruistic death in the cells of multicellular organisms such as mammals. Although yeast is a unicellular organism, many apoptotic factors have recently been identified in yeast, and several stimuli have been reported to induce mitochondria-mediated apoptosis in this organism. Ethanol-induced death in yeast also exhibits features of apoptosis mediated by the mitochondrial fission pathway under certain conditions. Whether yeast causes apoptosis during fermentation remains to be determined. However, it appears that manipulation of apoptotic factors in yeast has the potential to increase fermentation efficiency and to ensure the survival of yeast during storage.

Chapter 2 - Two of the main plants currently being considered as potential biofuel feedstocks in the U.S. are switchgrass (*Panicum virgatum* L.) and maize (*Zea mays* L.). Recent expanded production of both has raised serious questions about natural resource utilization, notably, soil carbon, soil nutrients, and water. Water is often the limiting resource for crop and grass productivity. The objective of this study was to calculate and compare water use and water use efficiency of maize with current growth characteristics, switchgrass with current growth characteristics, and switchgrass with characteristics improved by normal plant breeding selection techniques. We used the calibrated and validated ALMANAC model for five sites representing the southern Great Plains (Stephenville, TX), the northern Great Plains (Mead, NE), and two locations in the Corn Belt (Ames, IA and Columbia, MO). Ten years of historical weather data were used. Mean values for water use and water use efficiency were calculated for maize, switchgrass with currently growth characteristics, and switchgrass with anticipated improved growth characteristics. These results show the relative impact of expanded maize production, expanded switchgrass production, and use of improved switchgrass varieties, on the water balance in these regions. The water use efficiency (WUE) of four switchgrass types showed means ranging from 3 to 5 mg g^{-1}. Switchgrass WUE values were much greater than WUE of maize grain, but such was not always the case when compared to WUE of maize plants. Changes in switchgrass light extinction coefficients (k) and in switchgrass radiation use efficiency (RUE) showed the expected trends. Increased RUE caused increases in dry matter yield and in WUE, but not usually as great as the percentage increase in RUE. Results from this simulation work will give guidance to policy planners, producers, and economists.

Chapter 3 - Energy remains the mainstay for the entire civilized world. The concept dates back to 1885 when Rudolf Diesel built the first diesel engine with the full intention of running it on vegetative source. The Kyoto protocol has prompted resurgence in the use of biodiesel throughout the world. There is a growing interest in *Jatropha curcas* as a biodiesel 'miracle tree' to help alleviate the energy crisis, reduce the countries dependence on foreign oil imports and generate income in rural areas of developing countries. It is becoming a poster child amongst some proponents of renewable energy. *J.curcas* also called the physic nut is used to produce the non-edible *Jatropha* oil and the estimates of the oil content in seeds range from 35-40% and in the kernels 55-60%. India is the 5th largest energy consumer in the world. The country imported 90MT of crude oil by 2003-04 that was only 70 % of the requirement. By 2030 the estimated consumption is to scale up to 5.6 m barrels/day of which 95% is to be met by import. The ever-increasing demand can only be met by an alternative source as biofuel. Normal conventional propagation of *Jatropha curcas* has several drawbacks like poor seed viability, low germination, scanty and delayed rooting of seedlings, etc. The yield of nuts is unsure and unsustainable, coupled with unknown genetic potential. The plants propagated by cuttings show a lower longevity and possess a lower drought and disease resistance. Therefore, the need of the hour is to shake hands with biotechnological processes to overcome the hindrances. This chapter is just a hand forward to uniform, genetically stable, quality planting material yield, intricately interwoven with sustainability in energy. Plant biotechnology serves as a powerful tool for fast and quality plant production, and regerneration of plants in *J.curcas* has been successfully applied. Protocol development is the prerequisite for creating genetically improved crops, and three protocols have been successfully developed that will serve as an ideal system for future transgenic research and can act as a powerful tool for genetic improvement of the plant species. Various authors have reported *in vitro* micropropagation in *J.curcas*, using different explants. This chapter will highlight an up-to-date overview of the present and future trends of research in *Jatropha curcas* along with the work done in the author's laboratory.

The authors have reported micropropagation through nodal meristem culture from field grown plant. This is the first report of complete plant regeneration through somatic embryogenesis in this species. Somatic embryogenesis is the focus of all applied research and it is now considered as the gateway to many more technologies. Plant propagation by somatic embryogenesis not only helps to obtain a large number of plants year round, but also can act as a powerful tool for further transgenic research. In this review we are reporting complete plant production from excised immature zygotic embryos. This method can serve as a support system to conventional and modern agriculture for the genetic improvement of the crop.

Chapter 4 - On an experimental field in eastern Germany 10 haulm-type and woody crop species which are adequate for combustion and gasification have been cultivated under practical conditions on a sandy brown soil for 14 years now. Each crop received 4 different levels of fertilization, from 0 kg N ha^{-1} to 150 kg N ha^{-1} combined with straw and wood ashes. The measuring program includes yields, energy gain and environmentally relevant substances in plants and soil, as well as fertilizer-induced emissions. Measured long-term yields are between 7 and 10 t ha^{-1} y^{-1} for all crops with the exception of topinambur and a special poplar variety. In contrast to the yields of haulm-type crops which decrease along with reduced fertilization, the woody crops such as poplar and willow show an increase. Energy yields average out between 100 and 160 GJ ha^{-1} y^{-1}, while the energy demand for cultivation and harvesting only accounts for 1% to 13% of these yields.

The nitrogen content (N$_t$) varies between the crops in a range from 0.1% to 3.3% and depends on the fertilization level. With 0.2 to 1.2% N$_t$, poplars and willows have only half the N$_t$ of whole crop cereals and hemp, and approximately only one third that of cocksfoot grass. Therefore, these species cause less NO$_x$ emissions during combustion. Nitrous oxide (dinitrogen oxide: N$_2$O) flux measurements, carried out over nine years, show that the mean nitrogen conversion factor, which describes the N$_2$O greenhouse burden of fertilization, is about 0.8±0.1 % and thus slightly lower than the default value of 1% for N$_2$O inventories. Therefore, the production of lingo-cellulosic energy crop species considered here will not lose its CO$_2$ advantage by nitrogen fertilizing as long as fertilizing results in an adequately higher biomass yield.

Soil organic carbon (C$_{org}$) stocks are likely to be improved by a land-use change from annual to woody crops. Twelve years after establishment of the plantation, a comparison of C$_{org}$ stocks under annual crops with C$_{org}$ under short rotation willow and poplar revealed an increase of soil C$_{org}$ under the trees of 1300 kg ha^{-1} y^{-1}. By comparison, fertilization only accounts for a difference in soil C$_{org}$ between fertilized and non-fertilized tree blocks of 250 kg ha^{-1} y^{-1}.

Chapter 5 - Biomass is a key feedstock to produce renewable biofuels such as Fischer-Tropsch hydrocarbons, methanol, and hydrogen. However, limitation of land and water, and competition with food production reduce potential significance of biomass as a renewable energy source. The development of efficient conversion technologies, which are able to compete with fossil fuels, is a key challenge for biomass-based systems.

Production of biofuels is traditionally analyzed by energetic analysis based on the First Law of Thermodynamics. However, this type of analysis shows only the mass and energy flows and does not take into account how the quality of the energy and material streams degrades through the process. In this chapter the exergy analysis, which is based on the Second Law of Thermodynamics, is used to analyze the conversion of biomass to biofuels.

The most promising biomass-to-biofuel route is a two-stage process involving production of syngas from biomass gasification, followed by synthesis of transportation fuels. Several overall technological chains biomass-to-biofuels are evaluated, including Fischer-Tropsch hydrocarbons, hydrogen, and methanol. It is shown that that exergetic efficiency of production of biofuels is lower than that for fossil fuels.

Biomass gasification shows the largest exergy losses in the overall chain biomass-to-biofuels. Modification of process conditions to improve efficiency of gasification is discussed. It is shown that the optimal gasification conditions correspond to the carbon boundary point where all carbon present in biomass is gasified. The efficiency of biomass gasification can be also improved in a thermal pretreatment called torrefaction.

Chapter 6 - In this study energy balance and fuel properties of biodiesel has been calculated. Accordingly, the cost of 1 liter of oil is calculated 0.32 € after the income from the seed meal is deduced. Finally, the cost of per unit of biodiesel (1 liter) was calculated as 0.55 €, after deduction of the income provided by the sales of glycerin for use in soap and cosmetic industry.

The energy equivalent of total output was calculated 147605.50 MJ per hectare. The net energy gain (refined oil) was found as 15105.63 MJ per hectare (The net energy ratio 11.031) according to yield and inputs values.

The viscosity values of vegetable oils vary between 27.2 and 53.6 mm^2/s whereas those of vegetable oil methyl esters between 3.59 and 4.63 mm^2/s. The flash point values of vegetable oil methyl esters are highly lower than those of vegetable oils. The flash point values of vegetable oil methyl esters are highly lower than those of vegetable oils. An increase in density from 860 to 885 kg/m^3 for vegetable oil methyl esters or biodiesel increases the viscosity from 3.59 to 4.63 mm^2/s and the increases are highly regular. There is high regression between density and viscosity values vegetable oil methyl esters. The relationships between viscosity and flash point for vegetable oil methyl esters are irregular. An increase in density from 860 to 885 kg/m^3 for vegetable oil methyl esters increases the flash point from 401 to 453 K and the increases are slightly regular.

The LHV values of vegetable oils methyl ester vary between 35.74 and 39.16 MJ/kg.

Expert Commentary - Large scale production and consumption of biofuel is being promoted by many nations because of its potential for economic, political, and environmental benefits. Biofuels are partially renewable and substituting them for petroleum fuels may reduce emissions of greenhouse gases and other pollutants, and dependency on imported oil. Of special interest is the possibility of widespread use of biofuels such as biodiesel in the transportation sector, especially if integrated with existing and new hybrid and plug-in technologies and emission control technologies such as particle traps.

In: New Research on Biofuels   ISBN 978-1-60456-828-8
Editors: J. H. Wright and D. A. Evans   © 2008 Nova Science Publishers, Inc.

Chapter 1

THE MECHANISM OF ETHANOL-INDUCED DEATH IN YEAST AND ITS POTENTIAL APPLICATION FOR THE FERMENTATION INDUSTRY

Hiroshi Kitagaki

National Research Institute of Brewing, Senior Researcher
Hiroshima Prefecture, JAPAN

ABSTRACT

The production of ethanol by yeast is a key technology in ethanol fermentation. However, ethanol is toxic to yeast cells. It causes misfolding of proteins, an increase in membrane fluidity, changes in mRNA export from the nucleus, and activation of various stress signaling pathways, including the protein kinase A pathway. Ultimately, it causes cell death. Ethanol-induced death in yeast occurs during the fermentation of several alcoholic beverages such as sake (Japanese rice wine) and wine. In contrast, during most of the fermentation of bioethanol yeast cells do not die, either because, the ethanol concentration does not reach a high enough level, or because the fermentation period is not long enough to cause the yeast to die. However, future bioethanol production technology may require more severe fermentation conditions. Such technology could include the production of ethanol to a higher ethanol concentration, repetitive fermentation by fixed yeast cells, or fermentation at high temperature. Thus, ethanol-induced death during fermentation may also be a limiting factor in the future production of bioethanol. Therefore, elucidation of the mechanism of ethanol-induced death merits further research. This knowledge could be used for the development of technology to enhance yeast productivity in the fermentation of alcoholic beverages and bioethanol.

Cell death can be grouped into two categories—necrosis and apoptosis. While necrosis is a catastrophic and passive death, apoptosis is a programmed death that is of benefit to the organism as a whole. Historically, apoptosis was first identified as an altruistic death in the cells of multicellular organisms such as mammals. Although yeast is a unicellular organism, many apoptotic factors have recently been identified in yeast, and several stimuli have been reported to induce mitochondria-mediated apoptosis in this organism. Ethanol-induced death in yeast also exhibits features of apoptosis mediated by the mitochondrial fission pathway under certain conditions. Whether yeast causes apoptosis during fermentation remains to be determined. However, it appears that manipulation of apoptotic factors in yeast has the potential to increase fermentation efficiency and to ensure the survival of yeast during storage.

INTRODUCTION

Many fermentation industries utilize the ability of the yeast *Saccharomyces* to ferment and to produce ethanol. However, ethanol itself accumulates in yeast cells (D'Amore et al., 1988; Dombek and Ingram, 1996), is toxic, and ultimately causes the death of the yeast cells during some forms of ethanol fermentation. One such example is in the brewing of sake, Japanese rice wine. During sake brewing, the concentration of ethanol reaches as high as 21% volume per volume, which causes the yeast to die. During the production of bioethanol, the concentration of ethanol also reaches a toxic level for yeast cells. However, in many cases, the concentration of ethanol is not high enough, or the fermentation period is not long enough to cause death. However, future advances in fermentation technology of bioethanol might require more severe growth conditions for yeast cells. For example, if a higher temperature was used for faster fermentation, or if repeated fermentation was used in order to bypass propagation process of yeast cells. Therefore, yeast cell death during fermentation of bioethanol could be a problem in the future. For all of the above reasons, elucidation of the pathway by which ethanol induces cell death is potentially useful for enhancing fermentation productivity. This pathway could be used as a basis for the design of strategies to prevent the ethanol-induced death of yeast during industrial fermentation.

CELLULAR TARGETS OF ETHANOL TOXICITY IN YEAST

Ethanol has various deleterious effects on yeast cells, which are summarized in Figure 1. Although the precise mechanism by which ethanol induces cells death is complex and remains undefined, the primary targets of ethanol stress appear to be protein structure and membrane lipids. Examples of proteins whose structures are modulated by ethanol include GABA (A) and NMDA receptors of neurons (Peoples and Weight, 1999). Moreover, ethanol inhibits enzymes such as hexokinase (Augustin et al., 1986) and dehydrogenases (Nagodawithana and Steinkraus, 1976). One of the heat shock proteins, Hsp104, which can refold and reactivate aggregated proteins, has been shown to play a critical role in ethanol tolerance (Fahrenkrog et al., 2004), further suggesting a critical role for protein structure in the action of ethanol. Secondly, cell membranes are considered to be one of the primary targets of ethanol stress (Dombek and Ingram, 1984). Ethanol is believed to increase the membrane fluidity of yeast cells and induce oxygen-derived free radical attack of membrane lipids (Beaven et al., 1982; Alexandre et al., 1994a; Alexandre et al., 1994b; Chi and Arneborg, 1999; Ingram, 1976; Kajiwara et al., 1996; Mishra and Prasad, 1989; Swan and Watson, 1999). Consistent with this hypothesis, when yeast cells are supplemented with fatty acids, oleic acid ($C_{18:1}$) is the most potent in endowing cells with ethanol tolerance, while linoleic ($C_{18:2}$) or linolenic ($C_{18:3}$) acids have weaker effects on ethanol tolerance (Swan and

Watson, 1999). Moreover, You et al. demonstrated that ethanol tolerance is dependent on oleic acid content. They showed that a desaturase-deficient *ole1* knockout strain is deficient in growth in the presence of ethanol but the same cells, when supplemented with oleic acids, are rescued from this growth deficiency in ethanol-containing medium (You et al., 2003). From these data, they concluded that ethanol tolerance in yeast results from incorporation of oleic acid into lipid membranes, which leads to a compensatory decrease in membrane fluidity that counteracts the fluidity-increasing effect of ethanol. Other studies also confirm the role of fatty acids in ethanol resistance. Overexpression of *OLE1*, a desaturase gene (Kajiwara et al., 2000), or exogenous *FAD2*, a delta-12 fatty acid desaturase gene (Kajiwara et al., 1996), renders yeast cells tolerant to ethanol (*S. cerevisiae* does not encode *FAD2* in its genome). In addition, growth in a culture medium rich in palmitic acid ($C_{16:0}$) renders yeast cells ethanol tolerant (Ohta and Hayashida, 1983). Ergosterol, a rigid structural component of membrane lipids, also confers ethanol tolerance to yeast cells (Novotný et al., 1992; Inoue et al., 2000; Alexandre et al., 1994). Hsp12 is reported to be induced upon ethanol treatment, and to protect cell membranes (Sales et al., 2000). In support of this role of Hsp12, cells deleted in *HSP12* were extremely sensitive to ethanol (Sales et al., 2000). Growing yeast in a media containing lipids such as a mixture of a surfactant Tween 80, ergosterol and monoolein, or phosphatidylcholine (Hayashida et al., 1974; Hayashida et al., 1975; Hayashida et al., 1976; Hayashida and Ohta, 1978; Hayashida and Ohta, 1980; Ohta and Hayashida, 1983) increased the final ethanol concentration obtained following fermentation of sake yeast strains, indicating a critical role of these lipids in ethanol fermentation ability. From the observation that yeast cells enriched with phosphatidylserine had greater ethanol tolerance, it was suggested that the anion:zwitterion ratio of phospholipids may be one of the important determinants of ethanol tolerance in *S. cerevisiae* (Mishra and Prasad, 1998). The ratio of unsaturated fatty acids has been shown to increase at later stages of ethanol fermentation (Gil et al., 1990). There is also a study which reports that yeast cells adapted to increased concentrations of produced ethanol contain increased amount of ergosterol and decreased amount of lanosterol, increased amount of phosphatidylinositol and decreased amount of phosphatidylcholine, and increased amount of C18:0 fatty acids in total phospholipids and decreased amount of C16:0 fatty acids (Arneborg et al., 1995). It is also reported that palmitoyl-CoA pool and incorporation of exogenously added palmitic acid by Faa4 is critical for growth in the existence of ethanol (Nozawa et al., 2002). Several studies have reported that ethanol damages mitochondrial DNA (Castrejón et al., 2002; Ibeas and Jimenez, 1997). Moreover, ethanol treatment of yeast cells has been reported to increase reactive oxygen species, most of which reside in the mitochondria (Du and Takagi, 2007). Together with the observation that cells deficient in superoxide dismutase are sensitive to ethanol (Costa et al., 1997), these facts suggest a critical role of reactive oxygen species in the ethanol-response of yeast. More recently, it has been shown that leakage of reactive oxygen species is caused by membrane damage to the mitochondria caused by ethanol (Chi and Arneborg, 1999). The combined results are consistent with the hypothesis that ethanol causes membrane damage by enhancing membrane fluidity and that a reduction in membrane fluidity renders yeast cells tolerant to ethanol.

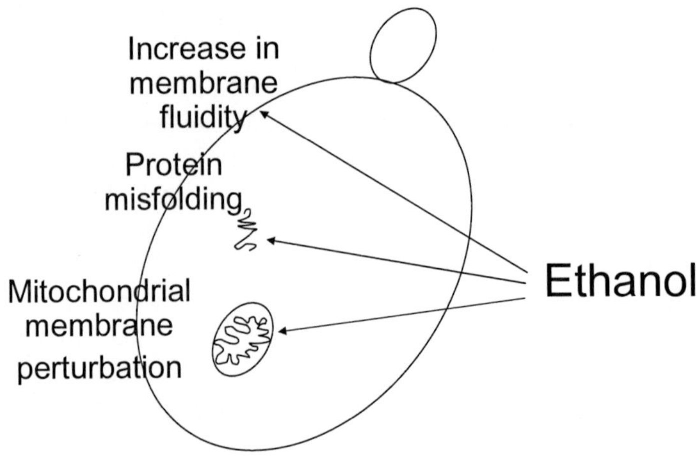

Figure 1. Cellular targets of ethanol toxicity in yeast.

PHYSIOLOGICAL RESPONSES OF YEAST CELLS CHALLENGED WITH ETHANOL

As a result of the toxic effects of ethanol on yeast cells described above, yeast cells induce several physiological responses (Figure 2). These responses include activation of cell signaling pathways and induction of cell-protective molecules such as heat shock proteins and trehalose. Signaling pathways activated by ethanol included the stress-responsive protein kinase A-Msn2 pathway (van Voorst et al., 2006; Martínez-Pastor et al., 1996). Ethanol treatment induces the translocation of a GFP-tagged-Msn2 into the nucleus (van Voorst et al., 2006; Watanabe et al., 2007). Stress response elements (STRE)-regulated genes such as *CTT1*, *DDR2* and *HSP12*, are also upregulated upon ethanol treatment (Martínez-Pastor et al., 1996; Schüller et al., 1994). It has been reported that one family of cell integrity and stress response sensors localized on the plasma membrane plays critical roles in the ethanol response of yeast (Zu et al., 2001). Yeast cells disrupted in these sensors (*wsc1-3*△) were sensitive to, and showed insufficient adaptation to ethanol. Wsc1-3 have been reported to activate the signaling cascade of Rho1, MAPK cascade (Bck1, Mkk1/2, Mpk1), Pkc1 and Rlm1 (Philip et al., 2001). Consistent with the involvement of this signaling pathway with ethanol tolerance, Takahashi reported that *BEM2* and *ROM2*, which encode a GAP (GTPase activating protein for Rho1) and a GEF (GDP-GTP exchange factor for Rho1), respectively, are required for the growth of yeast in ethanol (Takahashi et al., 2001). The combined data indicate that the Wsc1-3-Rho1-MAPK cascade-Pkc1-Rlm1 signaling pathway plays a critical role in ethanol tolerance.

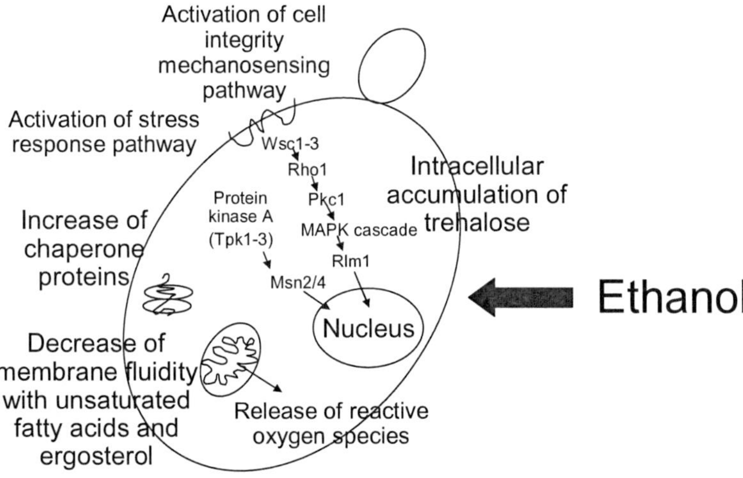

Figure 2. Physiological response of yeast cells challenged with ethanol.

Several metabolic constituents within yeast cells are required for ethanol tolerance. For example, intracellular trehalose protects yeast cells against ethanol inhibition of endocytosis (Lucero et al., 2000), its cellular level is increased upon ethanol stress (Sharma, 1997), and the trehalose biosynthesis genes are required for growth in the presence of ethanol (van Voorst et al., 2006; Kubota et al., 2004), although clear correlation between the trehalose content and viability could not be observed (Alexandre et al., 1998). Also, normal tryptophan biosynthesis is required for ethanol tolerance, since many yeast mutants that are deficient in tryptophan biosynthesis are ethanol sensitive, based on DNA microarray data analysis (Hirasawa et al., 2007). Inositol levels also influence ethanol tolerance, since increasing the inositol levels of yeast in static culture results in high ethanol tolerance, probably because of the resulting increase in phosphatidylinositol (Furukawa et al., 2004). Disruption of the genes *URA7* and *GAL6* also improves ethanol tolerance and the fermentation capacity of yeast, probably through modification of cell wall integrity and membrane fluidity (Yazawa et al., 2007).

Ethanol induces the transcription of specific sets of genes including genes involved in ionic homeostasis, heat protection, trehalose synthesis and antioxidant defense and energy metabolism (Alexandre et al., 2001). Moreover, ethanol stress also induces the accumulation of bulk polyA$^+$ mRNAs in the nucleus in a similar manner to that induced by heat shock (Takemura et al., 2004). However, the mRNAs of heat shock proteins are selectively exported from the nucleus whereas export of the DEAD box helicase Rat8p from the nucleus is observed only in ethanol-treated cells, but not in heat-shocked cells. This suggests that a mechanism of nuclear export exists that can distinguish between heat shock and ethanol shock. When yeast are grown on wine must and synthetic media the Rat8p accumulates in the nucleus as fermentation proceeds. However, during sake brewing, this protein only accumulates in the nucleus of the yeast until the ethanol concentration of the mash reaches approximately 12%. At a later stage of sake brewing, Rat8p re-localizes to the cytosol (Izawa et al., 2005a). The accumulation of bulk polyA$^+$ mRNAs in the nucleus has also been observed during wine brewing (Izawa et al., 2005b). These results suggest that selective export of polyA$^+$ mRNAs has a physiological role in wine fermentation and sake brewing.

Ethanol also enhances the formation of P-bodies, in which mRNAs, for example mRNAs that are not translated, are stored and degraded (Izawa et al., 2007). The formation of P-bodies was enhanced upon ethanol treatment and during fermentation of wine. In contrast, during fermentation of sake, the number and size of P-bodies began to increase gradually only after the ethanol concentration exceeded 13%.

Asr1, which is a ring/PHD protein that constitutively shuttles between the nucleus and the cytoplasm but accumulates in the nucleus upon exposure to ethanol, has been reported to be involved in ethanol sensing of yeast cells (Betz et al., 2004). The hypersensitivity of Asr1-deficient mutants to alcohol and sodium dodecyl sulfate could not be confirmed (Izawa et al., 2006). There is another study that reports the normal sensitivity of asr1 disruptant to 8% ethanol but the translocation of GFP-tagged Asr1 to the nucleus upon treatment with 6% ethanol (van Voorst et al., 2006). Therefore, the significance of Asr1 in ethanol sensing remains open to discussion.

APOPTOSIS AND ITS MECHANISM IN YEAST

There are two forms of cell death, necrosis and apoptosis. Necrosis is a catastrophic and passive death. Apoptosis is an altruistic cell suicide program first identified in the cells of multicellular organisms such as mammals. In multicellular organisms, the clear purpose of apoptosis is to sacrifice some cells for the benefit of the entire organism. Therefore, the reason why unicellular organisms might have an apoptotic program is not immediately apparent. One hypothesis is that unicellular organisms execute apoptosis in order to benefit siblings with similar genomes by release of nutrients from the apoptotic cells. However, nutrients emitted from apoptotic cells do not necessarily reach their siblings. A second hypothesis is based on the fact that apoptotic cells degrade their genomes (Ribeiro et al., 2006). Thus, it has been suggested that apoptotic cells release DNA fragments and that these DNA fragments cause homologous recombination in the neighboring cells. As a result, genes are transmitted from the apoptotic cells to their siblings. However, this hypothesis remains to be proved.

Many, but not all, of the known mammalian apoptotic factors are conserved in yeast. For example, caspase, which is the key protease of apoptosis in mammalian cells, is conserved in yeast and plays a critical role in apoptosis induced by H_2O_2 (Madeo et al., 2002), chronological aging (Herker et al., 2004), hyperosmotic stress (Silva et al., 2005) and viruses (Ivanovska et al., 2005) in yeast. However, it is clearly not involved in DNA damage-(Wysocki et al., 2004) or ethanol-induced apoptosis (Kitagaki et al., 2007a) in yeast. The enzymatic activity of the yeast caspase homolog was experimentally confirmed upon treatment of yeast with hydrogen peroxide (Madeo et al., 2002). *AIF1*, which encodes apoptosis-inducing factor, also regulates H_2O_2-induced apoptosis (Wissing et al., 2004). Aif1p was shown to translocate from the mitochondria to the nucleus upon treatment with an apoptotic stimulus and its in vitro DNA degradation activity has been demonstrated. Bir1p protein, the inhibitor of apoptosis protein, is conserved in yeast, is a substrate of the human pro-apoptotic serine protease Omi/HtrA2, and has been show to play a role in apoptotic events in yeast (Walter et al., 2006). Apoptotic nuclease, Tat-D, is also conserved in yeast and has been suggested to play a role in the yeast apoptotic process (Qiu et al., 2005). Nuclear serine protease HtrA-like protein was recently identified in yeast as Nma111p and was reported to function in apoptosis upon heat treatment and hydrogen peroxide treatment in yeast (Fahrenkrog et al., 2004). Endonuclease G, that is normally localized in the mitochondria, but translocates to the nucleus in diseased states in mammalian cells, is encoded by *NUC1* in yeast and has been shown to function in a manner similar to that in mammalian cells (Buttner et al., 2007). Nuc1p executes apoptosis independently of metacaspase, but its apoptotic activity is dependent on the permeability transition pore protein, karyopherin Kap123p, as well as on histone H2B. Histone chaperone *ASF1/CIA1* also plays a role in apoptosis in yeast (Yamaki et al., 2001). The release of cytochrome c from mitochondria is also considered to play a key role in acetic acid- and pheromone-induced apoptosis in yeast as it does in mammalian cells (Ludovico et al., 2002; Zhang et al., 2006;Yamaki et al., 2001). In spite of these components of apoptosis that are highly conserved between yeast and mammals, some mammalian apoptotic factors are not encoded by the yeast genome. These apoptotic genes not encoded for by yeast include BH3-only proteins that regulate the permeability of the outer membrane of mitochondria. However, to our surprise, some of these proteins, such as, Bax, Bcl-2, tBid, Bad and Puma, when transfected into yeast do function in apoptosis in yeast, although others, including BNip3, BNip3L and Noxa, do not (Kissová et al., 2006; Yang et al., 2006; Zheng et al., 2007; Gonzalvez et al., 2005; Li et al., 2005; Guscetti et al., 2005; Polcic et al., 2003;Tao et al., 1997). These BH3-only proteins appear to regulate reactive oxygen species, since Bcl-2 reverses survival defects in yeast caused by disruption of superoxide dismutase (Longo et al., 1997). It is possible that the Fis1 protein, that is a yeast mitochondrial fission protein, can function in a similar manner to Bcl-2 (Fannjiang et al., 2004). It has been reported that yeast cells lacking the Fis1 protein contain an increased amount of reactive oxygen species, indicating that Fis1 regulates the release of reactive oxygen from the mitochondria (Kitagaki et al., 2007a). Neither of the apoptotic proteins Smac or Diablo are conserved in yeast (Guscetti et al., 2005). These results imply that common and specific mechanisms underlie apoptosis in yeast and mammalian cells.

APOPTOTIC FEATURES IN ETHANOL-INDUCED DEATH IN YEAST

Several apoptotic-inducing stimuli have been described for yeast. These include aging (Laun et al., 2001), hydrogen peroxide (Madeo et al., 1999), acetic acid (Guaragnella et al., 2006; Ludovico et al., 2001), high osmotic stress (Silva et al., 2005), heat stress (Lee et al., 2007), NaCl (Wadskog et al., 2004), UV (Del Carratore et al., 2002), nitric oxide (Almeida et al., 2007), ammonia (Váchová et al., 2005), pheromones (Zhang et al., 2006; Pozniakovsky et al., 2004), aspirin (Sapienza et al., 2005), killer toxins (Reiter et al., 2005), a mutation in *CDC48* (Braun et al., 2006), an N-glycosylation defect (Hauptmann et al., 2006), decapping of mRNA (Mazzoni et al., 2003; Mazzoni et al., 2005), the formation of F-actin and hyperactivated Ras signaling (Gourlay and Ayscough, 2005a; Gourlay and Ayscough, 2005a; Gourlay and Ayscough, 2006; Gourlay et al., 2004) and defects in the initiation of DNA replication(Weinberger et al., 2005).

Recently, it has been shown that ethanol-induced death of yeast exhibits features of mitochondrial fission pathway-mediated apoptosis (Kitagaki et al., 2007a). Ethanol-killed yeast cells exhibited chromatin aggregation and fragmentation, DNA degradation, suppression of death by inhibition of de-novo protein synthesis as revealed by DAPI staining, TUNEL (terminal deoxynucleotidyl Transferase Biotin-dUTP Nick End Labeling) labeling of the nucleus, pulse field electrophoresis and cycloheximide treatment. Consistent with this result, reactive oxygen species have been shown to increase upon exposure to ethanol (Du and Takagi, 2007).The Yca1 or Aif1 is not apparently involved in ethanol-induced death, although inhibition of caspase activity rescued the ethanol-induced death of *fis1* $\triangle$ cells. When this data was combined with a comparison of the number of PI-positive and TUNEL-positive cells, the results suggested that not all of the cells die of apoptosis in ethanol-induced death, but that some of the cells die of necrosis. If all of the research on ethanol toxicity in yeast is combined, the following model is suggested. Ethanol treatment perturbs the mitochondrial membrane and, together with the modulation of protein structure, and the increase in membrane fluidity of other cellular membranes, induces yeast death by both apoptosis as well as necrosis. The ratio of apoptosis to necrosis appears to depend on the physiological state of the yeast cells at the time they are exposed to ethanol stress. These results suggest that inhibition of apoptosis during ethanol fermentation has the potential to at least partially enhance the survival of yeast cells challenged with ethanol, and, thereby to increase the productivity of ethanol fermentation to a certain extent.

Another study has shown that mitochondrial structure is present throughout the alcohol fermentation process and to fragment during the process (Figure 3, Kitagaki et al., 2007b). Since mitochondrial proteins have been reported to regulate (either accelerate or inhibit) apoptosis, this study suggests that introduction or manipulation of mitochondrial apoptotic factors such as BH3 family proteins (Ligr et al., 1998), cytochrome c (Ludovico et al., 2002), apoptosis-inducing factor (*AIF1*) (Wissing et al., 2004), Smac and Diablo (Anguiano-Hernandez et al., 2007) through genetic engineering techniques has the potential to endow yeast cells with the ability to encounter ethanol stress (Figure 4).

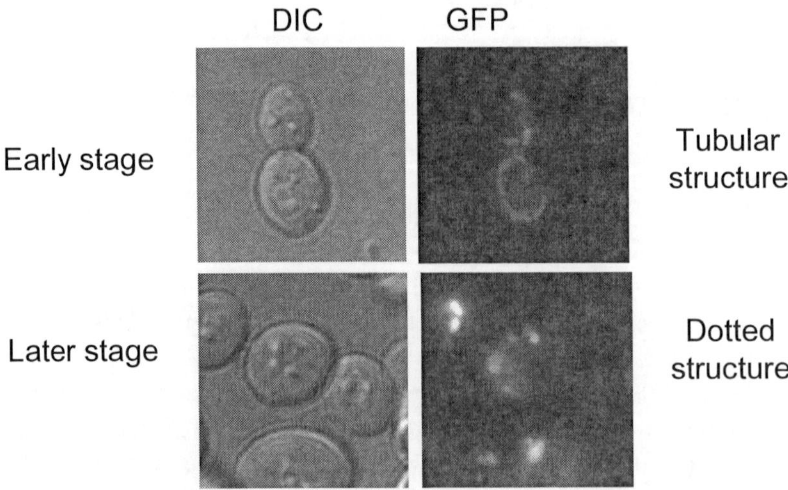

Figure 3. Mitochondrial morphology of yeast during alcohol fermentation.

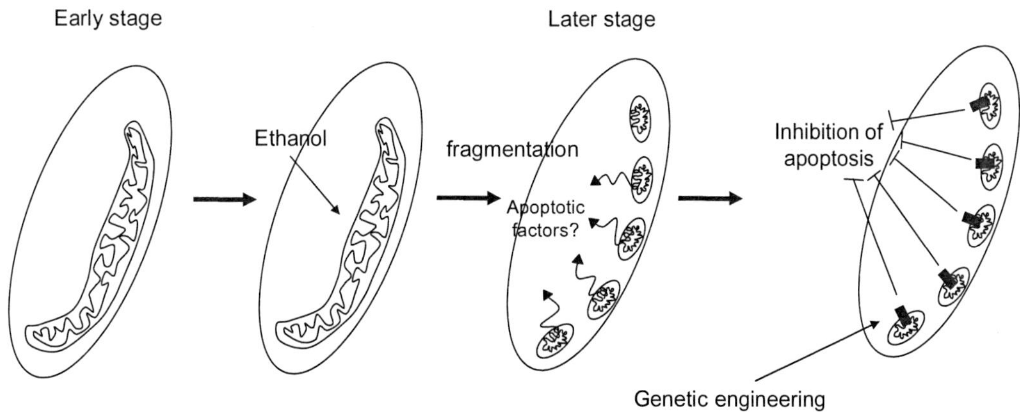

Figure 4. Future direction of breeding of ethanol-tolerant yeast strains for fermentation of bioethanol.

CONCLUSION

Although efforts have been made to increase the ethanol tolerance of yeast through the breeding of new yeast strains, manipulation of apoptotic pathways for this purpose has not yet been attempted. A combination of both of these approaches is likely to have a synergistic effect on the breeding of ethanol-tolerant yeast strains that would be of great value to the fermentation industry.

REFERENCES

Alexandre, H., Rousseaux, I., & Charpentier, C. (1994a). Ethanol adaptation mechanisms in *Saccharomyces cerevisiae. Biotechnol. Appl. Biochem. 20,* 173–183.

Alexandre, H., Rousseaux, I., & Charpentier, C. (1994b). Relationship between ethanol tolerance, lipid composition and plasma membrane fluidity in *Saccharomyces cerevisiae* and *Kloeckera apiculata.FEMS Microbiol. Lett. 124,* 17-22.

Alexandre, H., Plourde, L., Charpentier, C., & François, J. (1998). Lack of correlation between trehalose accumulation, cell viability and intracellular acidification as induced by various stresses in *Saccharomyces cerevisiae. Microbiology 144,* 1103-1111.

Alexandre, H., Ansanay-Galeote, V., Dequin, S., & Blondin, B. (2001). Global gene expression during short-term ethanol stress in *Saccharomyces cerevisiae. FEBS Lett. 498,* 98-103.

Almeida, B., Buttner, S., Ohlmeier, S., Silva, A., Mesquita, A., Sampaio-Marques, B., Osório, N. S., Kollau, A., Mayer, B., Leão, C., Laranjinha, J., Rodrigues, F., Madeo, F., & Ludovico, P. (2007). NO-mediated apoptosis in yeast. *J. Cell Sci. 120,* 3279-3288.

Anguiano-Hernandez, Y. M., Chartier, A., & Huerta, S. (2007). Smac/DIABLO and colon cancer. *Anticancer Agents Med. Chem. 7,* 467-473.

Arneborg, N., Høy, C. E., & Jørgensen, O. B. (1995). The effect of ethanol and specific growth rate on the lipid content and composition of *Saccharomyces cerevisiae* grown anaerobically in a chemostat. *Yeast 11,* 953-959.

Augustin, H. W., Kopperschlager, G., Steffen, H., & Hofmann, E. (1965). Hexokinase as limiting factor of anaerobic glucose consumption of *Saccharomyces carlsbergensis* NCYC74. *Biochim. Biophys. Acta 110,* 437–439.

Beaven, M. J., C. Charpentier, & A. H. Rose. (1982). Production and tolerance of ethanol in relation to phospholipid fatty-acyl composition in *Saccharomyces cerevisiae* NCYC. *J. Gen. Microbiol. 128,* 1447–1455.

Betz, C., Schlenstedt, G., & Bailer, S. M. (2004). Asr1p, a novel yeast ring/PHD finger protein, signals alcohol stress to the nucleus. *J. Biol. Chem. 279,* 28174-28881.

Braun, R. J., Zischka, H., Madeo, F., Eisenberg, T., Wissing, S., Büttner, S., Engelhardt, S. M., Büringer, D., & Ueffing, M. (2006). Crucial mitochondrial impairment upon CDC48 mutation in apoptotic yeast. *J. Biol. Chem. 281,* 25757-25767.

Büttner, S., Eisenberg, T., Carmona-Gutierrez, D., Ruli, D., Knauer, H., Ruckenstuhl, C., Sigrist, C., Wissing, S., Kollroser, M., Fröhlich, K. U., Sigrist, S., & Madeo, F. (2007). Endonuclease G regulates budding yeast life and death. *Mol. Cell 25,* 233-246.

Castrejón, F., Codón, A. C., Cubero, B., & Benítez, T. (2002). Acetaldehyde and ethanol are responsible for mitochondrial DNA (mtDNA) restriction fragment length polymorphism (RFLP) in flor yeasts. *Syst. Appl. Microbiol. 25,* 462-467.

Chi, Z., & Arneborg, N. (1999). Relationship between lipid composition, frequency of ethanol-induced respiratory deficient mutants, and ethanol tolerance in *Saccharomyces cerevisiae. J. Appl. Microbiol. 86,* 1047–1052.

Costa, V., Amorim, M. A., Reis, E., Quintanilha, A., & Moradas-Ferreira, P. (1997). Mitochondrial superoxide dismutase is essential for ethanol tolerance of *Saccharomyces cerevisiae* in the post-diauxic phase. *Microbiology 143,* 1649-1656.

D'Amore, T., Panchal, C. J., & Stewart, G. G. (1988). Intracellular ethanol accumulation in *Saccharomyces cerevisiae* during fermentation. *Appl. Environ. Microbiol. 54,* 110-114.

Del Carratore, R., Della Croce, C., Simili, M., Taccini, E., Scavuzzo, M., & Sbrana, S. (2002). Cell cycle and morphological alterations as indicative of apoptosis promoted by UV irradiation in *S. cerevisiae. Mutat. Res. 513,* 183-191.

Dombek, K. M., & Ingram, L. O. (1986). Determination of the intracellular concentration of ethanol in *Saccharomyces cerevisiae* during fermentation. *Appl. Environ. Microbiol. 51,* 197-200.

Dombek, K. M., & Ingram, L. O. (1984). Effects of ethanol on the *Escherichia coli* plasma membrane. *J. Bacteriol. 157,* 233-239.

Du, X., & Takagi, H. (2007). N-Acetyltransferase Mpr1 confers ethanol tolerance on *Saccharomyces cerevisiae* by reducing reactive oxygen species. *Appl. Microbiol. Biotechnol. 75,* 1343-1351.

Fahrenkrog, B., Sauder, U., & Aebi, U. (2004). The S. cerevisiae HtrA-like protein Nma111p is a nuclear serine protease that mediates yeast apoptosis. *J. Cell Sci. 117,* 115-126.

Fannjiang, Y., Cheng, W. C., Lee, S. J., Qi, B., Pevsner, J., McCaffery, J. M., Hill, R. B., Basañez, G., & Hardwick, J. M. (1994). Mitochondrial fission proteins regulate programmed cell death in yeast. *Genes Dev. 18,* 2785-2797.

Furukawa, K., Kitano, H., Mizoguchi, H., & Hara, S. (2004). Effect of cellular inositol content on ethanol tolerance of *Saccharomyces cerevisiae* in sake brewing. *J. Biosci. Bioeng. 98*, 107-113.

Gil, G. H., Jones, W. J., & Tornabene, T. G. (1990). Relationship of alcohol production to lipid composition of yeast in a continuous flow bioreactor. *Biochem. Cell Biol. 68*, 661-668.

Gonzalvez, F., Bessoule, J. J., Rocchiccioli, F., Manon, S., & Petit, P. X. (2005). Role of cardiolipin on tBid and tBid/Bax synergistic effects on yeast mitochondria.*Cell Death Differ. 12*, 659-667.

Guaragnella, N., Pereira, C., Sousa, M. J., Antonacci, L., Passarella, S., Côrte-Real, M., Marra, E., & Giannattasio, S. (2006). YCA1 participates in the acetic acid induced yeast programmed cell death also in a manner unrelated to its caspase-like activity. *FEBS Lett. 580*, 6880-6884.

Gourlay, C.W., Carpp, L.N., Timpson, P., Winder, S.J., & Ayscough, K.R. (2004). A role for the actin cytoskeleton in cell death and ageing in yeast. *J. Cell Biol. 164*, 803–809

Gourlay, C.W., & Ayscough, K. R. (2005a). The actin cytoskeleton: a key regulator of apoptosis and ageing. *Nat. Rev. Mol. Cell Biol. 6*, 583–589.

Gourlay, C.W., & Ayscough, K. R. (2005b). Identification ofan upstream regulatory pathway controlling actin-mediatedapoptosis in yeast. *J. Cell Sci. 118*, 2119–2132.

Gourlay, C.W., & Ayscough, K. R. (2006). Actin induced hyperactivation of the Ras signaling pathway leads to apoptosis in *S. cerevisiae*. *Mol. Cell Biol. 26*, 6487–6501.

Guscetti, F., Nath, N., & Denko, N. (2005). Functional characterization of human proapoptotic molecules in yeast *S. cerevisiae*. *FASEB J. 19*, 464-466.

Hauptmann, P., Riel, C., Kunz-Schughart, L. A., Fröhlich, K. U., Madeo, F., Lehle, L. 2006. Defects in N-glycosylation induce apoptosis in yeast. *Mol. Microbiol.* 59:765-778.

Hayashida, S., Feng, D. D., & Hongo, M. (1974). Function of the high concentration alcohol-producing factor. *Agric. Biol. Chem. 38*, 2001-2006.

Hayashida, S., Feng, D. D., & Hongo, M. (1975). Physiological properties of yeast cells grown in the proteolipidsupplemented media. *Agric. Biol. Chem. 39*, 1025-1031.

Hayashida, S., Feng, D. D., Ohta, K., Chaltiumvong, S., & Hongo, M. (1976). Compositions and a role of *Aspergillus oryzae*-proteolipid as a high concentration alcohol-producing factor. *Agric. Biol. Chem. 40*, 73-78.

Hayashida, S., & Ohta, K. (1978). Cell structure of yeasts grown anaerobically in Aspergillus oryzae-proteolipidsupplemented media. *Agric. Biol. Chem. 42*, 1139-1145.

Hayashida, S., & Ohta, K. (1980). Effects of phosphatidylcholine or ergosteryl oleate on physiological properties of *Saccharomyces sake*. *Agric. Biol. Chem. 44*, 2561-2567.

Herker, E., Jungwirth, H., Lehmann, K. A., Maldener, C., Fröhlich, K. U., Wissing, S., Büttner, S., Fehr, M., Sigrist, S., & Madeo, F. (2004). Chronological aging leads to apoptosis in yeast. *J. Cell Biol. 164*, 501-507.

Hirasawa, T., Yoshikawa, K., Nakakura, Y., Nagahisa, K., Furusawa, C., Katakura, Y., Shimizu, H., & Shioya, S. (2007). Identification of target genes conferring ethanol stress tolerance to *Saccharomyces cerevisiae* based on DNA microarray data analysis. *J. Biotechnol. 131*, 34-44.

Ibeas, J. I., & Jimenez, J. (1997). Mitochodrial DNA loss caused by ethanol in *Saccharomyces* flor yeasts. *Appl. Environ. Microbiol. 63*, 7–12.

Ingram, L. O. (1976). Adaptation of membrane lipids to alcohols. *J. Bacteriol. 125*, 670–678.

Inoue, T., Iefuji, H., Fujii, T., Soga, H., & Satoh, K. (2000). Cloning and characterization of a gene complementing the mutation of an ethanol-sensitive mutant of sake yeast. *Biosci. Biotechnol. Biochem. 64,* 229-236.

Ivanovska, I., & Hardwick, J. M. (2005). Viruses activate a genetically conserved cell death pathway in a unicellular organism. *J. Cell Biol. 170,* 391-399.

Izawa, S., Kita, T., Ikeda, K., Miki, T., & Inoue, Y. (2007). Formation of cytoplasmic P-bodies in sake yeast during Japanese sake brewing and wine making. *Biosci. Biotechnol. Biochem. 71,* 2800-2807

Izawa, S., Ikeda, K., Kita, T., & Inoue, Y. (2006). Asr1, an alcohol-responsive factor of *Saccharomyces cerevisiae,* is dispensable for alcoholic fermentation. *Appl. Microbiol. Biotechnol. 72,* 560-565.

Izawa, S., Takemura, R., Ikeda, K., Fukuda, K., Wakai, Y., & Inoue, Y. (2005a). Characterization of Rat8 localization and mRNA export in *Saccharomyces cerevisiae* during the brewing of Japanese sake. *Appl. Microbiol. Biotechnol. 69,* 86-91.

Izawa, S., Takemura, R., Miki, T., & Inoue, Y. (2005b). Characterization of the export of bulk poly(A)+ mRNA in *Saccharomyces cerevisiae* during the wine-making process. *Appl. Environ. Microbiol. 71,* 2179-2182.

Kajiwara, S., Shirai, A. Fujii, T., Toguri, T., Nakamura, K., & Ohtaguchi, K. (1996). Polyunsaturated fatty acid biosynthesis in *Saccharomyces cerevisiae*: expression of ethanol tolerance and the *FAD2* gene from *Arabidopsis thaliana. Appl. Environ. Microbiol. 62,* 4309–4313.

Kajiwara, S., Aritomi, T., Suga, K., Ohtaguchi, K., & Kobayashi, O. (2000). Overexpression of the *OLE1* gene enhances ethanol fermentation by *Saccharomyces cerevisiae. Appl. Microbiol. Biotechnol. 53,* 568-574.

Khan, M. A., Chock, P. B., & Stadtman, E. R. (2005). Knockout of caspase-like gene, *YCA1,* abrogates apoptosis and elevates oxidized proteins in *Saccharomyces cerevisiae. Proc. Natl. Acad. Sci. U S A. 102,* 17326-17331.

Kissová, I., Plamondon, L. T., Brisson, L., Priault, M., Renouf, V., Schaeffer, J., Camougrand, N., & Manon, S. (2006). Evaluation of the roles of apoptosis, autophagy, and mitophagy in the loss of plating efficiency induced by Bax expression in yeast. *J. Biol. Chem. 281,* 36187-36197.

Kitagaki, H., Araki, Y., Funato, K., & Shimoi, H. (2007). Ethanol-induced death in yeast exhibits features of apoptosis mediated by mitochondrial fission pathway. *FEBS Lett. 581,* 2935-2942.

Kitagaki, H., & Shimoi H. (2007). Mitochondrial dynamics of yeast during sake brewing. *J. Biosci. Bioeng. 104,* 227-230.

Kubota, S., Takeo, I., Kume, K., Kanai, M., Shitamukai, A., Mizunuma, M., Miyakawa, T., Shimoi, H., Iefuji, H., & Hirata, D. (2004). Effect of ethanol on cell growth of budding yeast: genes that are important for cell growth in the presence of ethanol. *Biosci. Biotechnol. Biochem. 68,* 968-972.

Laun, P., Pichova, A., Madeo, F., Fuchs, J., Ellinger, A., Kohlwein, S., Dawes, I., Fröhlich, K. U., & Breitenbach, M. (2001). Aged mother cells of *Saccharomyces cerevisiae* show markers of oxidative stress and apoptosis. *Mol. Microbiol. 39,* 1166-1173.

Lee, Y. J., Hoe, K. L., & Maeng, P. J. (2007). Yeast cells lacking the *CIT1*-encoded mitochondrial citrate synthase are hypersusceptible to heat- or aging-induced apoptosis. *Mol. Biol. Cell 18,* 3556-3567.

Li, A., & Harris, D. A. (2005). Mammalian prion protein suppresses Bax-induced cell death in yeast. *J. Biol. Chem. 280,* 17430-17434.

Ligr, M., Madeo, F., Fröhlich, E., Hilt, W., Fröhlich, K. U., & Wolf, D. H. (1998). Mammalian Bax triggers apoptotic changes in yeast. *FEBS Lett. 438,* 61-65.

Longo, V. D., Ellerby, L. M., Bredesen, D. E., Valentine, J. S., & Gralla, E. B. (1997). Human Bcl-2 reverses survival defects in yeast lacking superoxide dismutase and delays death of wild-type yeast. *J. Cell Biol. 137,* 1581-1588.

Ludovico, P., Sousa, M. J., Silva, M. T., Leão, C., & Côrte-Real, M. (2001). *Saccharomyces cerevisiae* commits to a programmed cell death process in response to acetic acid. *Microbiology 147,* 2409-2415.

Ludovico, P., Rodrigues, F., Almeida, A., Silva, M. T., Barrientos, A., & Côrte-Real, M. (2002). Cytochrome c release and mitochondria involvement in programmed cell death induced by acetic acid in *Saccharomyces cerevisiae. Mol. Biol. Cell 13,* 2598-2606.

Lucero, P., Peñalver, E., Moreno, E., & Lagunas, R. (2000). Internal trehalose protects endocytosis from inhibition by ethanol in *Saccharomyces cerevisiae. Appl. Environ. Microbiol. 66,* 4456-4461.

Madeo, F., Fröhlich, E., Ligr, M., Grey, M., Sigrist, S. J., Wolf, D. H., & Fröhlich, K. U. (1999). Oxygen stress: a regulator of apoptosis in yeast. *J. Cell Biol. 145,* 757-767.

Madeo, F., Herker, E., Maldener, C., Wissing, S., Lächelt, S., Herlan, M., Fehr, M., Lauber, K., Sigrist, S. J., Wesselborg, S., & Fröhlich, K. U. (2002). A caspase-related protease regulates apoptosis in yeast. *Mol. Cell 9,* 911-917.

Martínez-Pastor, M. T., Marchler, G., Schüller, C., Marchler-Bauer, A., Ruis, H., & Estruch, F. (1996). The *Saccharomyces cerevisiae* zinc finger proteins Msn2p and Msn4p are required for transcriptional induction through the stress response element (STRE). *EMBO J. 15,* 2227-2235.

Mazzoni, C., Herker, E., Palermo, V., Jungwirth, H., Eisenberg, T., Madeo, F., & Falcone, C. (2005). Yeast caspase 1 links messenger RNA stability to apoptosis in yeast. *EMBO Rep. 6,* 1076-1081.

Mazzoni, C., Mancini, P., Verdone, L., Madeo, F., Serafini, A., Herker, E., & Falcone, C. (2003). A truncated form of KlLsm4p and the absence of factors involved in mRNA decapping trigger apoptosis in yeast. *Mol. Biol. Cell. 14,* 721-729.

Mishra, P., & Prasad R. (1988). Role of phospholipid head groups in ethanol tolerance of *Saccharomyces cerevisiae. J. Gen. Microbiol. 134,* 3205-3211.

Mishra, P., & R. Prasad. (1989). Relationship between ethanol tolerance and fatty acyl composition of *Saccharomyces cerevisiae. Appl. Microbiol.Biotechnol. 30,* 294-298.

Nagodawithana, T. W., & Steinkraus, K. H. (1976). Influence of the rate of ethanol production and accumulation on the viability of *Saccharomyces cerevisiae* in rapid fermentation. *Appl. Environ. Microbiol. 31,* 158-162.

Novotný, C., Flieger, M., Panos, J., & Karst, F. (1992). Effect of 5,7-unsaturated sterols on ethanol tolerance in *Saccharomyces cerevisiae. Biotechnol. Appl. Biochem. 15,* 314-320.

Nozawa, M., Takahashi, T., Hara, S., and Mizoguchi, H. (2002). A role of *Saccharomyces cerevisiae* fatty acid activation protein 4 in palmitoyl-CoA pool for growth in the presence of ethanol. *J. Biosci. Bioeng. 93,* 288-295.

Ohta, K., & Hayashida, S. (1983). Role of Tween 80 and Monoolein in a Lipid-Sterol-Protein Complex Which Enhances Ethanol Tolerance of Sake Yeasts. *Appl. Environ. Microbiol. 46,* 821-825.

Peoples, R. W., & Weight, F. F. (1999). Differential alcohol modulation of GABA(A) and NMDA receptors. *Neuroreport 10,* 97-101.

Philip, B., & Levin, D. E. (2001). Wsc1 and Mid2 are cell surface sensors for cell wall integrity signaling that act through Rom2, a guanine nucleotide exchange factor for Rho1. *Mol. Cell Biol. 21,* 271-280.

Polcic, P., & Forte, M. (2003). Response of yeast to the regulated expression of proteins in the Bcl-2 family. *Biochem. J. 374,* 393-402.

Pozniakovsky, A. I., Knorre, D. A., Markova, O. V., Hyman, A. A., Skulachev, V. P., & Severin, F. F. (2005). Role of mitochondria in the pheromone- and amiodarone-induced programmed death of yeast. *J. Cell Biol. 168,* 257-269.

Qiu, J., Yoon, J. H., & Shen, B. (2005). Search for apoptotic nucleases in yeast: role of Tat-D nuclease in apoptotic DNA degradation. *J. Biol. Chem. 280,* 15370-15379.

Reiter, J., Herker, E., Madeo, F., & Schmitt, M. J. (2005). Viral killer toxins induce caspase-mediated apoptosis in yeast. *J. Cell Biol. 168,* 353-358.

Ribeiro, G. F., Côrte-Real, M., & Johansson, B. (2006). Characterization of DNA damage in yeast apoptosis induced by hydrogen peroxide, acetic acid, and hyperosmotic shock. *Mol. Biol. Cell 17,* 4584-4591.

Sales, K., Brandt, W., Rumbak, E., & Lindsey, G. (2000). The LEA-like protein HSP12 in *Saccharomyces cerevisiae* has a plasma membrane location and protects membranes against desiccation and ethanol-induced stress. *Biochim. Biophys. Acta. 1463,* 267-278.

Sanchez, Y., Taulien, J., Borkovich, K. A., & Lindquist, S. (1992). Hsp104 is required for tolerance to many forms of stress. *EMBO J. 11,* 2357-2364.

Sapienza, K., & Balzan, R. (2005). Metabolic aspects of aspirin-induced apoptosis in yeast. *FEMS Yeast Res. 5,* 1207-1213.

Schüller, C., Brewster, J. L., Alexander, M. R., Gustin, M. C., & Ruis, H. (1994). The HOG pathway controls osmotic regulation of transcription via the stress response element (STRE) of the *Saccharomyces cerevisiae CTT1* gene. *EMBO J. 13,* 4382-4389.

Sharma, S. C. (1997). A possible role of trehalose in osmotolerance and ethanol tolerance in *Saccharomyces cerevisiae. FEMS Microbiol. Lett. 152,* 11-15.

Silva, R. D., Sotoca, R., Johansson, B., Ludovico, P., Sansonetty, F., Silva, M. T., Peinado, J. M., & Côrte-Real, M. (2005). Hyperosmotic stress induces metacaspase- and mitochondria-dependent apoptosis in *Saccharomyces cerevisiae. Mol. Microbiol. 58,* 824-834.

Swan, T. M., & Watson, K. (1999). Stress tolerance in a yeast lipid mutant: membrane lipids influence tolerance to heat and ethanol independently of heat shock proteins and trehalose. *Can. J. Microbiol. 45,* 472–479.

Takahashi, T., Shimoi, H., & Ito, K. (2001). Identification of genes required for growth under ethanol stress using transposon mutagenesis in *Saccharomyces cerevisiae. Mol. Genet. Genomics. 265,* 1112-1119.

Takemura, R., Inoue, Y., & Izawa, S. (2004). Stress response in yeast mRNA export factor: reversible changes in Rat8p localization are caused by ethanol stress but not heat shock. *J. Cell Sci. 117,* 4189-4197.

Tao, W., Kurschner, C., & Morgan, J. I. (1997). Modulation of cell death in yeast by the Bcl-2 family of proteins. *J. Biol. Chem. 272,* 15547-15552.

Váchová, L., & Palková, Z. (2005). Physiological regulation of yeast cell death in multicellular colonies is triggered by ammonia. *J. Cell Biol. 169,* 711-717.

van Voorst, F., Houghton-Larsen, J., Jonson, L., Kielland-Brandt, M.C. & Brandt, A. (2006). Genome-wide identification of genes required for growth of *Saccharomyces cerevisiae* under ethanol stress. *Yeast 23,* 351–359.

Wadskog, I., Maldener, C., Proksch, A., Madeo, F., & Adler, L. (2004). Yeast lacking the SRO7/SOP1-encoded tumor suppressor homologue show increased susceptibility to apoptosis-like cell death on exposure to NaCl stress. *Mol. Biol. Cell 15,* 1436-1444.

Walter, D., Wissing, S., Madeo, F., & Fahrenkrog, B. (2006). The inhibitor-of-apoptosis protein Bir1p protects against apoptosis in *S. cerevisiae* and is a substrate for the yeast homologue of Omi/HtrA2. *J. Cell Sci. 119,* 1843-1851.

Watanabe, M., Tamura, K., Magbanua, J. P., Takano, K., Kitamoto, K., Kitagaki, H., Akao, T., & Shimoi, H. (2007). Elevated expression of genes under the control of stress response element (STRE) and Msn2p in an ethanol-tolerance sake yeast Kyokai no. 11. *J. Biosci. Bioeng. 104,* 163-170.

Weinberger, M., Ramachandran, L., Feng, L., Sharma, K., Sun, X., Marchetti, M., Huberman, J. A., & Burhans, W. C. (2005). Apoptosis in budding yeast caused by defects in initiation of DNA replication. *J. Cell Sci. 118,* 3543-3553.

Wissing, S., Ludovico, P., Herker, E., Büttner, S., Engelhardt, S. M., Decker, T., Link, A., Proksch, A., Rodrigues, F., Corte-Real, M., Fröhlich, K. U., Manns, J., Candé, C., Sigrist, S. J., Kroemer, G., & Madeo, F. (2004). An AIF orthologue regulates apoptosis in yeast. *J. Cell Biol. 166,* 969-974.

Wysocki, R., & Kron, S. J. (2004). Yeast cell death during DNA damage arrest is independent of caspase or reactive oxygen species. *J. Cell Biol. 166,* 311-316.

Yang, Z., Khoury, C., Jean-Baptiste, G., & Greenwood, M. T. (2006). Identification of mouse sphingomyelin synthase 1 as a suppressor of Bax-mediated cell death in yeast. *FEMS Yeast Res. 6,* 751-762.

Yamaki, M., Umehara, T., Chimura, T., & Horikoshi, M. (2001). Cell death with predominant apoptotic features in *Saccharomyces cerevisiae* mediated by deletion of the histone chaperone *ASF1/CIA1*. *Genes Cells 6,* 1043-1054.

Yazawa, H., Iwahashi, H., & Uemura, H. (2007). Disruption of *URA7* and *GAL6* improves the ethanol tolerance and fermentation capacity of *Saccharomyces cerevisiae*. *Yeast 24,* 551-560.

You, K. M., Rosenfield, C.L. & Knipple, D.C. (2003) Ethanol tolerance in the yeast *Saccharomyces cerevisiae* is dependent on cellular oleic acid content. *Appl. Environ. Microbiol. 69,* 1499–1503.

Zhang, N. N., Dudgeon, D. D., Paliwal, S., Levchenko, A., Grote, E., & Cunningham, K. W. (2006) Multiple signaling pathways regulate yeast cell death during the response to mating pheromones. *Mol. Biol. Cell 17,* 3409-3422.

Zheng, K., Pan, J. W., Ye, L., Fu, Y., Peng, H. Z., Wan, B. Y., Gu, Q., Bian, H. W., Han, N., Wang, J. H., Kang, B., Pan, J. H., Shao, H. H., Wang, W. Z., & Zhu, M. Y. (2007) Programmed cell death-involved aluminum toxicity in yeast alleviated by antiapoptotic members with decreased calcium signals. *Plant Physiol. 143,* 38-49.

Zu, T., Verna, J., & Ballester, R. (2001). Mutations in WSC genes for putative stress receptors result in sensitivity to multiple stress conditions and impairment of Rlm1-dependent gene expression in *Saccharomyces cerevisiae*. *Mol. Genet. Genomics 266,* 142-155.

In: New Research on Biofuels
Editors: J. H. Wright and D. A. Evans

ISBN 978-1-60456-828-8
© 2008 Nova Science Publishers, Inc.

Chapter 2

BIOFUELS AND WATER USE: COMPARISON OF MAIZE AND SWITCHGRASS AND GENERAL PERSPECTIVES

J. R. Kiniry[1], Lee Lynd[2], Nathanel Greene[3], Mari-Vaughn V. Johnson[1], Michael Casler[4] and Mark S. Laser[2]

[1] USDA-ARS, Grassland Res. Center, 808 East Blackland Road, Temple, TX 76502, USA
[2] Dartmouth University, Hanover, NH 03755-8000, USA
[3] Natural Resources Defense Council, New York, NY 10011, USA
[4] USDA-ARS, Madison, WI 53706, USA

ABSTRACT

Two of the main plants currently being considered as potential biofuel feedstocks in the U.S. are switchgrass (*Panicum virgatum* L.) and maize (*Zea mays* L.). Recent expanded production of both has raised serious questions about natural resource utilization, notably, soil carbon, soil nutrients, and water. Water is often the limiting resource for crop and grass productivity. The objective of this study was to calculate and compare water use and water use efficiency of maize with current growth characteristics, switchgrass with current growth characteristics, and switchgrass with characteristics improved by normal plant breeding selection techniques. We used the calibrated and validated ALMANAC model for five sites representing the southern Great Plains (Stephenville, TX), the northern Great Plains (Mead, NE), and two locations in the Corn Belt (Ames, IA and Columbia, MO). Ten years of historical weather data were used. Mean values for water use and water use efficiency were calculated for maize, switchgrass with currently growth characteristics, and switchgrass with anticipated improved growth characteristics. These results show the relative impact of expanded maize production, expanded switchgrass production, and use of improved switchgrass varieties, on the water balance in these regions. The water use efficiency (WUE) of four switchgrass types showed means ranging from 3 to 5 mg g^{-1}. Switchgrass WUE values were much greater than WUE of maize grain, but such was not always the case when compared to WUE of maize plants. Changes in switchgrass light extinction coefficients

(k) and in switchgrass radiation use efficiency (RUE) showed the expected trends. Increased RUE caused increases in dry matter yield and in WUE, but not usually as great as the percentage increase in RUE. Results from this simulation work will give guidance to policy planners, producers, and economists.

INTRODUCTION

Two of the plant species of of major interest for biofuel production in the U.S. are maize (*Zea mays* L.) and switchgrass (*Panicum virgatum* L.). Widespread expansion of the production area of these two species has raised concerns regarding competition of biofuel production with food and fiber production as well as concern for competition over resources needed for biofuel, food, and fiber production.

Three major resources needed for production of biomass are light, nutrients, and water. Light is virtually nonlimiting during the growing season. Nutrient requirements have become a key concern as the price of inorganic fertilizers continues to rise and concerns regarding runoff-pollution continue to grow. Finally, water is an especially important resource as the competition for limited water supply becomes more intense in this country. Decisions on water allocation have to be made regarding the production of food, fiber, and fuel.

Direct measurement of water use requires labor-intensive procedures involving soil water measurement with neutron access tubes, gravimetric measurements of soil moisture with soil cores, or use of weighing lysimeters. Likewise, measurements of WUE require plant harvesting to determine dry weight of whole plants or of grain. To adequately define WUE over a range of soils, plant species cover, and climatic conditions with such measurements would require an exorbitant amount of resources and time.

The best alternative to such extensive, expensive projects is to build a process-based simulation model, using data from field experimentation to develop the equations for water use and for plant growth. Once validated, this model could be applied to a diversity of soils, weather data, and plant species. This approach lead to the development of models such as EPIC (originally the Erosion Productivity Impact Calculator, now the Environmental Policy Integrated Climate) (Williams et al., 1984), SWAT (Soil and Water Assessment Tool) (Arnold et al., 1998), and ALMANAC (Agricultural Land Management Alternatives with Numerical Assessment Criteria) (Kiniry et al., 1992). These models simulate the water balance using various methods of calculating potential evapotranspiration and determine plant water use while considering such variables as soils, weather, and plant species cover. Simulated plant growth is reduced when simulated soil water is depleted. The plant growth model simulates changes in leaf area as well as changes in plant biomass and grain (in the case of crops).

Water use efficiency is thereby calculated with these models as dry weight of plant biomass (or grain) produced per unit water transpired (EP) or per unit total evapotranspiration (ET). In this study, we used the ALMANAC model to calculate WUE of maize, currently available switchgrass types, and anticipated improved switchgrass types in four locations in the central U.S. These locations are representative sites in the region anticipated to be the primary production area for biofuel crops in the U.S.

METHODS

Calculation of Water Use Efficiency

Water use efficiency (WUE) has been calculated both in terms of assimilation rate per unit water transpired (Nippert et al., 2007) or as plant dry weight increase per unit water used. In the latter case, plant dry weight can be either total above-ground dry weight or grain dry weight (for crops). Water use can be the amount of water transpired by plants during the growth period or total water lost (evapotranspiration) from the area, including both plant transpiration and soil evapotranspiration. For the present study, we calculated WUE as plant dry weight increase per unit water transpired. For maize, we also calculated WUE as grain weight increase per unit water transpired.

Nippert et al. (2007) reported values of WUE of 1.0 mmol CO_2 per mol of water for big bluestem (*Andropogon gerardii* Vitman) and 2.6 for indiangrass (*Sorghastrum nutans* (L.) Nash) in the field. Taking molecular weights into account, these values are 1.7 and 4.3 mg of CH_2O per g of water transpired. In an outdoor pot experiment, WUE values of different cultivars of switchgrass ranged from 4.3 to 8.5 mg dry weight per g of water used (Byrd and May, 2000). In germplasm nurseries in Tennessee and Oklahoma, WUE of switchgrass accessions were 3.5 to 6.3 mg CH_2O per g of water transpired (McLaughlin et al., 2006).

In direct terms of crop plant mass increase per unit water transpired, values of 1 to 5 mg dry weight per g of water are common (Hatfield et al., 2001). For maize above-ground dry matter, the WUE values at Bushland, TX were 1.66 to 2.34 mg g^{-1} (Eck and Winter, 1992), 2.75 to 2.88 mg g^{-1} (Howell et al., 1998), and 2.34 to 3.06 mg g^{-1} (Tolk et al., 1998). For maize grain yield, WUE values at Bushland were 0.24 to 0.81 (Eck and Winter, 1992), 1.52 to 1.57 mg g^{-1} (Howell et al., 1998), and 1.05 to 1.63 mg g^{-1} (Tolk et al., 1998). Elsewhere, WUE values for maize grain yield were 0.58 to 1.15 mg g^{-1} in Mead, NE (Varvel, 1994), 0.61 to 1.66 in Lexington, KY (Corak et al., 1991), and 1.9 to 2.3 mg g^{-1} in Iowa (Hatfield et al., 2001).

Grass WUE values also range between 1 to 5 mg dry matter production per g of water transpired. Blue grama (*Bouteloua gracilis* (H.B.K.)) in a greenhouse had a WUE value of 4.55 mg g^{-1} (Fairbourn, 1982). In a greenhouse, grass seedlings (*Sporobolus arabicus* and *Leptochloa fusca*) had WUE values of 1.0 to 1.4 mg g^{-1} (Akhter et al., 2003). In the field in Nebraska, switchgrass WUE values were 1.0 to 5.5 mg g^{-1} (Eggemeyer et al., 2006), values similar to those demonstrated by switchgrass seedlings in a growth chamber (1.45 to 5.5 mg g^{-1}; Xu et al., 2006). In the shortgrass steppe of Colorado, a mixture of cool-season and warm-season grasses (including blue grama) had WUE values of 1.0 to 4.5 mg g^{-1} (Nelson et al., 2004).

General Model Description

The ALMANAC model has been described numerous times as it has been used to simulate crops (Kiniry et al., 1997; Kiniry and Bockholt, 1998; Yun Xie et al., 2001) and warm season grasses (Kiniry et al., 1996; Kiniry et al., 2002; Kiniry et al., 2005; Kiniry et al., 2007; McLaughlin et al., 2006). Parameters to simulate different plants continue to be refined as new research results are reported. Basically, the model simulates the soil water balance, the soil and plant nutrient balance, and the interception of solar radiation. This model includes subroutines and functions from the EPIC model (Williams et al., 1984, 1990) with added details for plant growth. The model has a daily time step. It simulates plant growth for a wide range of species and is implemented easily.

Light Interception

ALMANAC simulates light interception by the leaf canopy with Beer's law (Monsi and Saeki, 1953) and the leaf area index (LAI). The LAI is the amount of leaf area per unit ground area, a unitless variable. With greater extinction coefficient values (k), a given LAI intercepts more light.

The fraction of incoming solar radiation intercepted by the leaf canopy is

$$FRACTION = 1.0 - exp(k \times LAI) \tag{1}$$

Leaf Area Development

Accurate prediction of light interception depends on realistic simulation of leaf area. The model estimates leaf area production up to the point of maximum leaf area for the growing season using Eq. [2]. The sigmoid-curve function for potential LAI production takes the form:

$$F = SYP/[SYP + exp(Y1 - Y2 \times SYP)] \tag{2}$$

Where F is the factor for relative LAI, SYP is the fraction of the degree days from planting to maturity, and Y1 and Y2 are the sigmoid-curve coefficients generated by ALMANAC. This curve passes through the origin and through two points, asymptotically approaching F = 1.0. The model calculates SYP each day. The sum of degree days is zero at planting in the establishment year and at tiller emergence in subsequent years, and reaches its maximum value at maturity.

The model describes the loss of leaf area late in the season with the LAI decline factor. The LAI begins to decrease after a defined fraction of the seasonal degree days have accumulated.

Biomass Production and Partitioning

The model simulates biomass with an RUE value for each plant species (Kiniry et al., 1989). Values for RUE have a wide range of crops and grasses (Kiniry et al., 1989; Kiniry et al., 1992; Kiniry et al., 1999; Kiniry et al., 2007). ALMANAC describes declining RUE in later growth stages with an identical function to the one for the decrease in LAI.

The maximum rooting depth defines the potential depth in the absence of a root-restricting soil layer. Soil cores at Temple, TX in 1994 showed that switchgrass roots extend to depths of at least 2.2 m.

Water and Nutrient Uptake

Critical for yield and biomass simulation in water-limited conditions is the simulated water demand. The ALMANAC model calculates effects of soil water on crop growth and yield with similar functions. Potential evaporation is calculated first, then potential soil water evaporation and potential plant water transpiration are derived from potential evaporation and leaf area index. Based on the soil water supply and crop water demand, the water stress factor is estimated to decrease daily crop growth and yield. However, some water balance equations differ between the two models. For this study, potential evaporation was estimated by the Penman-Monteith method (Ritchie, 1972). Potential soil evaporation and plant transpiration were estimated by:

$$E_P = E_0(LAI/3) \qquad\qquad 0 \leq LAI \leq 3.0 \qquad\qquad [3]$$

$$E_P = E_0 \qquad\qquad LAI > 3.0 \qquad\qquad [4]$$

$$E_S = \text{minimum of } (E_0 \exp(-0.1BIO), E_0\text{-}E_P) \qquad\qquad [5]$$

Where E_P and E_s are potential plant transpiration and soil evaporation (mm), E_0 is potential evaporation (mm), LAI is leaf area index, and BIO is the sum of the above ground biomass and crop residue (Mg ha^{-1}).

$$E_P = E_0 \qquad\qquad LAI > 3.0 \qquad\qquad [6]$$

$$E_S = E_0(1\text{-}0.43\, LAI) \qquad\qquad 0 \leq LAI \leq 1.0 \qquad\qquad [7]$$

$$E_S = E_0 \exp(-0.4LAI)/1.1 \qquad\qquad LAI > 1.0 \qquad\qquad [8]$$

Water stress factor (WSF) is the ratio of water use to water demand (potential plant transpiration) in ALMANAC, and water use (WU) is a function of plant extractable water and root depth.

The nutrient balance (N and P) also allows plants to acquire sufficient nutrients to meet the demands if adequate quantities are available in the current rooting zone. Nutrient values for switchgrass were refined with N concentration data collected at Stephenville during 5 years (Sanderson, unpublished data) and Waller et al. (1972).

Base Temperature, Optimum Temperature, and Total Degree Days

Base temperature in ALMANAC is constant for all growth stages. Base temperature constrains the initiation of leaf area growth and thus dry matter accumulation. Higher optimum temperature can allow increased plant development rate later in the season when temperatures are greater. The sum of degree days to maturity controls the duration of growth. Heat units are reset to zero after maturity each year. Heat units are calculated from daily maximum and minimum temperatures, assuming the maximum equals the optimum if it exceeds the optimum.

Parameters Used to Simulate Four Switchgrass Types

We used an extensive, published data set (Casler et al., 2004) with two years of measured data at five sites of varying latitude to derive and verify switchgrass parameters. The four switchgrass types were Northern Upland (NU), Northern Lowland (NL), Southern Upland (SU), and Southern Lowland (SL). The locations were Spooner, WI (42° 49′ N 91° 54′ W), Arlington, WI (43° 20′ N 89° 23′ W), Mead, NE (41° 13′ N 96° 29′ W), Manhattan, KS (39°12′ N 96° 15′ W) and Stillwater, OK (36° 07′ N 96° 05′ W). The soil type at Spooner was Omega loamy sand (soil depth=1.52 m, plant available water=6.7 cm). The soil type at Arlington was Plano silt loam (soil depth=1.83 m, plant available water=37.0 cm). The soil type at Mead was Sharpsburg silt loam (soil depth=1.52 m, plant available water=29.3 cm). The soil type at Manhattan was Plano silt loam (soil depth=1.52 m, plant available water=30.6 cm). The soil type at Stillwater was Kirkland silt loam (soil depth=2.03 m, plant available water=27.0 cm). Runoff curve number was set to 71 for all the locations. Plots were planted in spring, 1998. Plots were fertilized with 112 kg N ha^{-1} in spring each year. Plots were harvested and biomass was quantified in late summer 1999 and 2000.

Parameters were adjusted to get reasonable simulated yields of each switchgrass type, as compared to the mean of the two years of measured yields. Using parameters developed for Alamo switchgrass in Texas (Kiniry et al., 1996; Kiniry et al., 1999; and Kiniry et al., 2007), the only two parameters adjusted were the degree days to maturity (base 12 C) and the potential leaf area index (PotLAI). The light extinction coefficient for Beer's Law (Monsi and Saeki, 1953) was set to -0.51, the average of the two means from Kiniry et al. (1999) and Kiniry et al., (2007).

The values of PotLAI and degree days varied with latitude and switchgrass types (Table 1). Other parameters used for all the locations and switchgrass types were:

1. 3.9 g per MJ intercepted photosynthetically active radiation for the radiation use efficiency (RUE) at low vapor pressure deficits (VPD)
2. 0.65 unit decrease in RUE for each 1 kPa increase in VPD above 1 kPa
3. Base temperature of 12 C and optimum temperature of 25 C
4. Linear decreases in LAI and in RUE from 70% of total degree day accumulation until maturity
5. Maximum potential rooting depth of 2.2 m
6. Optimum nitrogen concentrations of 2.57% for plants early in the spring, 1.1% for plants near mid season, and 0.28% for plants near maturity each year
7. Optimum phosphorus concentrations of 0.14% for plants early in the spring, 0.10% for plants near mid season, and 0.07% for plants near maturity each year.

Table 1. Data sets (Casler, et al., 2004) to develop parameters for and validate the ALMANAC model for four switchgrass types

Switchgrass Type:	SL	NL	SU	NU
Stillwater, OK (36° N)				
Degree days	1200	1100	1050	950
Pot LAI	3.0	3.0	2.5	1.8
Msrd Yield	15.13	14.81	12.62	10.45
Sim. Yield	15.12	15.45	13.65	11.31
Sim./Msrd.	1.00	1.04	1.08	1.08
Overall mean Sim./Msrd=1.05				
Manhattan, KS (39° N)				
Degree days	1200	1100	1050	950
Pot LAI	3.0	3.0	2.5	2.0
Msrd Yield	9.26	10.28	7.96	7.05
Sim. Yield	9.14	8.62	8.17	6.86
Sim./Msrd.	0.99	0.84	1.03	0.97
Overall mean Sim./Msrd=0.96				
Mead, NE (41° N)				
Degree days	1100	1100	900	800
Pot LAI	3.0	3.0	2.5	2.0
Msrd Yield	17.46	20.93	14.99	12.71
Sim. Yield	16.84	17.75	15.30	12.92
Sim./Msrd.	0.96	0.85	1.03	1.03
Overall mean Sim./Msrd=0.97				
Arlington, WI (43° N)				
Degree days	800	1100	800	800
Pot LAI	1.8	2.4	2.5	2.0
Msrd Yield	6.46	10.61	11.44	10.25
Sim. Yield	7.07	10.70	11.56	9.88
Sim./Msrd.	1.09	1.01	1.01	1.02
Overall mean Sim./Msrd=1.02				
Spooner, WI (46 ° N)				
Degree days	540	630	900	800
Pot LAI	1.8	1.8	2.5	2.0
Msrd Yield	3.81	4.60	7.39	7.33
Sim. Yield	4.07	4.76	7.96	6.73
Sim./Msrd.	1.07	1.04	1.08	0.95
Overall mean Sim./Msrd=1.03				

The four types were Southern Lowland (SL), Northern Lowland (NL), Southern Upland (SU), and Northern Upland (NU). Degree days are the values (base 12 C) to maturity each growing season. Pot LAI is the input potential leaf area index. Msrd. Yield is the published yield (Mg ha^{-1}). Sim Yield is the yield (Mg ha^{-1}) simulated by ALMANAC. Sim/Msrd is the ratio of simulated divided by measured yields

We showed good overall agreement between simulated and measured yields following the adjustment of degree days and potential LAI (Table 1). For Stillwater, OK, simulated yields were within 8% of measured yields, with the mean simulated/measured ratio being 1.05. For Manhattan, KS, simulated yields were within 3% of measured for three of the four switchgrass types. The mean overall ratio of simulated/measured was 0.96. For Mead, NE, simulated yields were within 4% of measured for three of the four switchgrass types. The mean simulated/measured ratio was 0.97. For the two sites in WI, the simulated yields were always within 9% of the measured yields, with the two mean simulated/measured ratios being 1.02 and 1.03. Thus, overall, we were able to realistically simulate the measured yields of the four switchgrass types at these diverse latitudes by adjusting only two plant parameters.

Locations Simulated to Compare WUE of Four Switchgrass Types to WUE of Maize

Four sites were chosen to represent a major portion of the area in the U.S. potentially useful for biofuel production: Stephenville, TX (32° 13′ N 98° 13′ W), representative of the Southern Great Plains; Mead, NE (41° 13′ N 96° 29′ W), representative of the Northern Great Plains; and Columbia, MO (38° 57′ N 92° 19′ W) and Ames, IA (42° 1′ N 93° 42′ W), which represent the Corn Belt. We used 10 years of measured weather data and three representative soils for each of the pertinent counties (Table 2).

Table 2. Soils used for switchgrass and maize simulations

Soil Type	Depth	PAW	CN
	m	m	
Mead, NE			
Yutan silty clay loam	2.03	0.292	71
Tomek silt loam	1.83	0.256	71
Nodaway silt loam	1.52	0.222	71
Ames, IA			
Clarion loam	1.52	0.187	69
Nicollet loam	1.52	0.225	69
Webster clay loam	1.52	0.192	69
Columbia, MO			
Keswick silt loam	1.52	0.172	78
Mexico silt loam	1.52	0.167	78
Weller silt loam	1.52	0.205	78
Stephenville, TX			
Brackett clay loam	1.52	0.166	89
Altoga clay loam	1.68	0.215	89
Houston Black clay	2.00	0.284	89

Plant available water (PAW) is the difference between field capacity and wilting point. CN is the runoff curve number. The first soil listed for each site is the most common for the county and was used for the simulations with improved switchgrass types.

Parameters Adjusted to Simulate Improved Switchgrass Types

Two key plant characteristics that may be manipulated in efforts to improve switchgrass yields are 1. the mean leaf angle and 2. photosynthetic rate. We simulated different leaf angles by changing the light extinction coefficient from -0.25 (upright leaf types), to -0.50 (intermediate leaf angles), and to -0.75 (relatively flat leaf types). We simulated possible increases in photosynthetic rate by increasing the RUE by 5%, 10% and 20%.

Only one switchgrass type was used at each site for these simulations. The type was the Southern Lowland for Stephenville, TX and was the Northern Upland for the other three sites. Only the most common soil of the three for each site was used for these simulations.

RESULTS

The WUE values for switchgrass and maize were similar to those reported in the literature. Comparisons between switchgrass types and maize, for WUE resulted in differing results depending on the location and the switchgrass type. Likewise, yields and WUE of improved switchgrass types showed interesting variations among locations.

Table 3. Mean water use efficiency values (mg of biomass per g of water transpired) simulated by ALMANAC for 10 years

Ames, Iowa	SL	NL	SU	NU	Mplant	Mgrain
soil 1	4.1	4.0	4.5	3.3	3.7	2.2
soil 2	4.1	4.6	3.2	2.8	3.5	2.2
soil 3	4.5	4.3	3.6	2.8	3.5	2.2
Mead, Nebraska						
soil 1	5.3	5.4	3.9	3.6	4.3	2.0
soil 2	5.0	4.9	3.4	3.0	4.2	2.1
soil 3	5.0	4.9	3.4	3.0	4.4	2.0
Columbia, Missouri						
soil 1	4.5	4.6	4.1	3.9	5.6	2.5
soil 2	4.3	4.5	3.7	3.2	4.2	1.8
soil 3	4.2	4.3	3.7	3.2	4.2	1.9
Stephenville, Texas						
soil 1	3.5	3.3	3.2	3.2	4.2	1.4
soil 2	3.5	3.2	3.1	3.1	4.0	1.4
soil 3	3.5	3.2	3.2	3.1	3.9	1.6

There are four switchgrass types, Southern Lowland (SL), Northern Lowland (NL), Southern Upland (SU), and Northern Upland (NU) (Casler et al., 2004). Maize whole plant (Mplant) and maize grain (Mgrain) WUE values are included for comparison. Soils are in the same order as in Table 2.

Water Use Efficiency of Four Switchgrass Types

The WUE of all four switchgrass types showed means ranging from 3 to 5 mg g^{-1} (Table 3). The greatest WUE values were nearly always for the lowland types. The one exception was for the first soil in Ames. The northern lowland type had the highest WUE in more than half the cases in the three northern locations. In Texas, the southern lowland type had the highest WUE for all three soils. For comparison, the maize types had WUE values of 3.5 to 5.6 for plant WUE and 1.4 to 2.5 for grain WUE, similar to reported values in the literature.

Comparing Water Use Efficiency of Switchgrass to WUE of Maize

Switchgrass WUE values were much greater than WUE of maize grain, but such was not always the case when compared to WUE of maize plants (Table 4). When compared to WUE of maize grain, switchgrass WUE was 1.8 to 5.0 times as great. Thus switchgrass produces more biomass per unit water transpired than does maize grain, currently the most common source for ethanol in the U.S. When compared to maize plant biomass, switchgrass WUE was greater in all the northern three sites except for the second soil at Columbia. However, for the Stephenville site, maize biomass WUE was always greater than switchgrass WUE. Thus for a given amount of soil water, we simulated greater biomass yields for switchgrass than for maize biomass at the first three sites, but lower yields for switchgrass than for maize at the southernmost site, Stephenville, Texas.

Table 4. Ratios of water use efficiency values of lowland switchgrass (WUEsw) to water use efficiency of maize, both simulated by ALMANAC for 10 years. Comparison with WUE of maize grain (WUEgrain) and maize total above-ground biomass (WUEbiomass) are included. Soils are in the same order as in Table 2.

	soil 1	soil 2	soil 3
Ames, Iowa			
WUEsw/WUEgrain	2.08	1.85	1.90
WUEsw/WUEbiomass	1.28	1.10	1.17
Mead, Nebraska			
WUEsw/WUEgrain	2.70	5.00	2.43
WUEsw/WUEbiomass	1.24	1.17	1.14
Columbia, Missouri			
WUEsw/WUEgrain	2.42	1.84	2.18
WUEsw/WUEbiomass	1.02	0.80	1.01
Stephenville, Texas			
WUEsw/WUEgrain	2.42	2.57	3.50
WUEsw/WUEbiomass	0.87	0.85	0.90

Water Use Efficiency of Improved Switchgrass Types

Changes in switchgrass light extinction coefficients (k) and in switchgrass RUE showed the expected trends, but not necessarily the magnitude of responsiveness that we hypothesized (Table 5). Biomass yields consistently increased as the k increased from -0.25 to -0.50 and to -0.75. The increase was expected due to the greater light interception prior to leaf canopy closure, with increased k values. The largest increase, however, caused only a 2 to 5% increase in yield, relative to the original switchgrass parameters. Correspondingly, WUE increased as k increased, with highest values 2 to 7% greater than the WUE of the original switchgrass.

Table 5. For potentially improved switchgrass types, mean biomass yields (yield), mean transpiration (EP), mean evapotranspiration (ET), water use efficiency (WUE, mg of biomass per g of water transpired), WUE of the improved type as compared to the original switchgrass type (WUE/WUEorig), and ratio of biomass yield of improved type to biomass yield of original type (YIELD/YIELDorig)

	original	Ext25	Ext50	Ext75	RUE1.05	RUE1.1	RUE1.2
Ames, Iowa							
yield (Mg/ha)	17.23	11.68	17.19	18.09	17.46	17.68	18.08
EP (mm)	399	457	399	397	399	398	397
ET (mm)	669	728	670	665	669	667	666
WUE (mg per g)	4.32	2.56	4.31	4.56	4.38	4.44	4.55
WUE/WUEorig	1.00	0.59	1.00	1.06	1.01	1.03	1.05
YIELD/YIELDorig	1.00	0.68	1.00	1.05	1.01	1.03	1.05
Mead, Nebraska							
yield (Mg/ha)	15.65	9.2	15.58	16.03	15.83	15.9	16.03
EP (mm)	290	292	291	293	292	283	292
ET (mm)	623	652	623	619	622	621	619
WUE (mg per g)	5.39	3.15	5.36	5.47	5.42	5.62	5.48
WUE/WUEorig	1.00	0.59	0.99	1.02	1.01	1.04	1.02
YIELD/YIELDorig	1.00	0.59	1.00	1.02	1.01	1.02	1.02
Columbia, Missouri							
yield (Mg/ha)	15.85	10.63	15.81	16.29	16.02	16.12	16.27
EP (mm)	356	472	360	356	356	356	356
ET (mm)	681	759	686	668	677	675	675
WUE (mg per g)	4.46	2.25	4.39	4.58	4.50	4.53	4.57
WUE/WUEorig	1.00	0.50	0.98	1.03	1.01	1.02	1.02
YIELD/YIELDorig	1.00	0.67	1.00	1.03	1.01	1.02	1.03
Stephenville, Texas							
yield (Mg/ha)	13.91	9.27	13.84	14.64	15.28	15.58	16.08
EP (mm)	393	370	393	385	425	422	412
ET (mm)	656	674	656	652	663	611	651
WUE (mg per g)	3.54	2.51	3.52	3.80	3.59	3.69	3.90
WUE/WUEorig	1.00	0.71	0.99	1.07	1.01	1.04	1.10
YIELD/YIELDorig	1.00	0.67	0.99	1.05	1.10	1.12	1.16

These are the results of 10 years of simulation by the ALMANAC model on the most common soil for each site. Columns represent potential improvements to switchgrass. Ext25 assumes a light extinction coefficient value of -0.25. Ext50 assumes a light extinction value of -0.50. Ext75 assumes a light extinction coefficient value of -0.75. RUE1.05 assumes a 5% increase in RUE. RUE1.10 assumes a 10% increase in RUE. RUE 1.2 assumes a 20% increase in RUE

Increased RUE caused increases in dry matter yield and in WUE, but not usually as great as the percentage increase in RUE. Increases in RUE of 5% caused yields to increase 1% in the northern three locations, and by 12% in the southern location. Increases in RUE of 10% caused yields to increase by 2 to 3% in the northern three locations and by 12% in the southern location. However, there was a diminishing return; when RUE was doubled from 10% to 20%, yields only increased by 2-5% at the northern locations and by 16% in Texas. Correspondingly, WUE values increased by greater percentages at the southern location than at the northern three sites as RUE was increased.

DISCUSSION

As the U.S. and other countries continue to expand the use of alternative energy sources, bioenergy in the form of maize grain, maize biomass, and switchgrass biomass are being investigated. Their use raises numerous questions, not the least of which are how will widespread expansion of their production affect natural resources through changes in soil erosion, through changes in soil carbon, and through changes in water use. The latter becomes especially important as this county defines strategies to allocate its limited water to feed and clothe its population, as well as supply energy needs.

The results presented in this paper are useful guides for how efficiently maize and switchgrass use water to produce biomass or grain (for maize). Switchgrass is very efficient when compared with the commonly used source of ethanol, maize grain.

Likewise, these results give some guidance for best selection strategies for future breeding programs on improving switchgrass biomass and switchgrass WUE. The ALMANAC simulation model can be extremely valuable as a research tool to define where breeding efforts can best be applied for improving switchgrass for biofuel.

REFERENCES

Akhter, J., K. Mahmood, M.A. Tasneem, M.H. Naqvi & K.A. Malik. (2003) Comparative water-use efficiency of *Sporobolus arabicus* and *Leptochloa fusca* and its relation with carbon-isotope discrimination under semi-arid conditions . *Plant and Soil* 249(2): 263-269.

Arnold, J. G., Srinivasan, R., Muttiah, R. S., & Williams, J. R. (1998) Large area hydrologic modeling and assessment part I: model development. *Journal American Water Resources Association,* 34, 73-89.

Casler, M.D., Vogel, K. P., Taliagerro, C. M., & Wynia, R. L. (2004) Latitudinal adaptation of switchgrass populations. *Crop Science,* 44, 293-303.

Corak, S. J., Frye, W. W., & Smith, M.S. (1991) Legume mulch and nitrogen fertilizer effects on soil water and corn production. *Soil Science Society of America Journal,* 58, 1395-1400.

Eck, H. V., & Winter, S.R. (1992) Soil profile modification effects on corn and sugarbeet grown with limited water. *Soil Science Society of America Journal,* 56, 1298-1304.

Eggemeyer, K. D., Awada, T., Wedin, D. A., Harvey, F. E., & Xinhua Zhou. (2006) Ecophysiology of two native invasive woody species and two dominant warm-season grasses in the semiarid grassland of the Nebraska sandhills. *Internacional Journal Plant Science,* 167, 5, 991-999.

Fairbourn, M. L. (1982) Water use by forage species. *Agronomy Journal,* 74, 62-66.

Hatfield. J. L., Sauer, T. J., & Prueger, J. H. (2001) Managing soils to achieve greater water use efficiency: A review. *Agronomy Journal,* 93, 271-280.

Howell, T. A., Tolk, J. A., Schneider, A. D., & Evett, S. R. (1998) Evapotranspiration, yield, and water use efficiency of corn hybrids differing in maturity. *Agronomy Journal,* 90, 3-9.

Kiniry, J. R., Jones, C.A., O'Toole, J.C., Blanchet, R., Cabelguenne, M., & Spanel, D. A. (1989) Radiation-use efficiency in biomass accumulation prior to grain-filling for five grain-crop species. *Field Crops Research,* 20, 51-64.

Kiniry J. R., Williams, J. R., Gassman, P. W., & Debaeke, P. (1992) A general process-oriented model for two competing plant species. *Transactions of American Society of Agricultural Engineers,* 35, 801-810.

Kiniry, J. R., Sanderson, M. A., Williams, J. R., Tischler, C. R., Hussey, M. A., Ocumpaugh, W. R., Read, J. C., VanEsbroeck, G., & Reed, R. L. (1996) Simulating Alamo switchgrass with the ALMANAC model. *Agron J.* 88:602-606.

Kiniry, J. R., Williams, J. R., Vanderlip, R. L., Atwood, J. D., Reicosky, D. C., Mulliken, J., Cox, W. J., Mascagni,H. J. Jr., Hollinger, S. E., & Wiebold, W. J. (1997) Evaluation of two maize models for nine U.S. locations. *Agronomy Journal,* 89, 421–426.

Kiniry, J.R., & Bockholt, A. J. (1998) Maize and sorghum simulation in diverse Texas environments. *Agronomy Journal,* 90, 682-687.

Kiniry, J.R., Tischler, C.R., & Van Esbroeck, G.A. (1999) Radiation use efficiency and leaf CO_2 exchange for diverse C_4 grasses. *Biomass and Bioenergy,* 17, 95-112.

Kiniry, J. R., Sanchez, H., Greenwade, J., Seidensticker, E., Bell, J. R., Pringle, F., Peacock, G. Jr., & Rives, J. (2002) Simulating grass productivity on diverse range sites in Texas. *Journal of Soil and Water Conservation,* 57, 144-150.

Kiniry, J. R., Cassida, K. A., Hussey, M. A., Muir, J. P., Ocumpaugh W. R., Read, J. C., Reed, R. L., Sanderson, M. A., Venuto, B. C., & Williams, J. R. (2005) *Switchgrass simulation by the ALMANAC model at diverse sites in the southern U.S. Biomass & Bioenergy* 29, 419-425.

Kiniry, J. R., Burson, B. L., Evers, G. W., Williams, J. R., Sanchez, H., Wade, C., Featherston, J. W., & Greenwade, J. (2007) Coastal bermudagrass, bahiagrass, and native range simulation at diverse sites in Texas. *Agronomy Journal* 99, 450-461.

McLaughlin, S. B., Kiniry, J. R., Taliaferro, C. M., & De LaTorre Ugarte, D. (2006) Projecting yield and utilization potential of switchgrass as an energy crop. *Advances in Agronomy,* 90, 267-297.

Monsi, M., & Saeki, T. (1953) Über den lichtfaktor in den pflanzengesellschaften und seine bedeutung fur die stoffproduktion. *Japan Journal of Botany,* 14, 22-52.

Nelson, J. A., Morgna, J. A., LeCain, D. R., Mosier, A. R., Milchunas, D. G., & Parton, B. A. (2004) Elevated CO_2 increases soil moisture and enhances plant water relations in a long-term study in semi-arid shortgrass steppe of Colorado. *Plant and Soil,* 259, 169-179.

Nippert, J. B., Fay, P. A., & Knapp, A. K. (2007) Photosynthetic traits in C3 and C4 grassland species in mesocosm and field environments. *Environmental and Experimental Botany,* 60, 412-420.

Ritchie, J. T. (1972) Model for predicting evaporation from a row crop with incomplete cover. *Water Resources Research,* 8, 1204-1213.

Tolk, J. A., Howell, T. A., & Evett, S. R. (1998) Evapotranspiration and yield of corn grown on three High Plains soils. *Agronomy Journal,* 90, 447-454.

Varvel, G. E. (1994) Monoculture and rotation system effects on precipitation use efficiency of corn. *Agronomy Journal,* 86, 204-208.

Waller, G. R., Morrison, R. D., & Nelson, A. B. (1972) Chemical composition of native grasses in central Oklahoma from 1947 to 1962. *Oklahoma Agricultural Experiment Station Bulletin* B-697.

Williams, J. R., Jones, C. A., & Dyke, P. T. (1984) A modeling approach to determining the relationship between erosion and soil productivity. *Transactions of American Society of Agricultural Engineers,* 27, 129-144.

Williams, J. R., Dyke, P. T., Fuchs, W. W., Benson, V. W., Rice, O. W., & Taylor, E. D. (1990) EPIC- erosion/productivity impact calculator: 2. User manual. In: A. N. Sharpley & J. R. Williams, editors, *USDA Technical Bulletin No. 1768.* 127 pp.

Xie, Yun, Kiniry, J. R., Nedbalek, V., & Rosenthal, W. D. (2001) Maize and sorghum simulations with CERES-Maize, SORKAM, and ALMANAC under water-limiting conditions. *Agronomy Journal,* 93, 1148-1155.

Xu, B., Li, F., Shan, L., Ma, Y., Ichizen, N., & Huang, J. (2006) Gas exchange, biomass partition, and water relationships of three grass seedlings under water stress. *Weed Biology and Management,* 6, 79-88.

In: New Research on Biofuels
Editors: J. H. Wright and D. A. Evans

ISBN 978-1-60456-828-8
© 2008 Nova Science Publishers, Inc.

Chapter 3

BIOTECHNOLOGICAL IMPROVEMENT OF A BIOFUEL CROP: *JATROPHA CURCAS* LINN.

Priyanka Mukherjee and Timir Baran Jha[*]
Department of Botany, Presidency College, Kolkata -700 073
West Bengal, India

ABSTRACT

Energy remains the mainstay for the entire civilized world. The concept dates back to 1885 when Rudolf Diesel built the first diesel engine with the full intention of running it on vegetative source. The Kyoto protocol has prompted resurgence in the use of biodiesel throughout the world. There is a growing interest in *Jatropha curcas* as a biodiesel 'miracle tree' to help alleviate the energy crisis, reduce the countries dependence on foreign oil imports and generate income in rural areas of developing countries. It is becoming a poster child amongst some proponents of renewable energy. *J.curcas* also called the physic nut is used to produce the non-edible *Jatropha* oil and the estimates of the oil content in seeds range from 35-40% and in the kernels 55-60%. India is the 5th largest energy consumer in the world. The country imported 90MT of crude oil by 2003-04 that was only 70 % of the requirement. By 2030 the estimated consumption is to scale up to 5.6 m barrels/day of which 95% is to be met by import. The ever-increasing demand can only be met by an alternative source as biofuel. Normal conventional propagation of *Jatropha curcas* has several drawbacks like poor seed viability, low germination, scanty and delayed rooting of seedlings etc. The yield of nuts is unsure and unsustainable, coupled with unknown genetic potential. The plants propagated by cuttings show a lower longevity and possess a lower drought and disease resistance. Therefore the need of the hour is to shake hands with biotechnological processes to overcome the hindrances. This chapter is just a hand forward to uniform, genetically stable, quality planting material yield, intricately interwoven with sustainability in energy. Plant biotechnology serves as a powerful tool for fast and quality plant production and we have been successful to apply it to regenerate plants in *J.curcas*. Protocol development is the prerequisite for creating genetically improved crops and we

---

[*] Corresponding author. E-mail i.d: presibot@vsnl.net,tbjha2000@yahoo.co.in

have successfully developed three protocols that will serve as an ideal system for future transgenic research and can act as a powerful tool for genetic improvement of the plant species. Various authors have reported *in vitro* micropropagation in *J.curcas*, using different explants. This chapter will highlight an up-to-date overview of the present and future trends of research in *Jatropha curcas* along with the work done in the author's laboratory.

We have reported micropropagation through nodal meristem culture from field grown plant. Ours is the first report of complete plant regeneration through somatic embryogenesis in this species. Somatic embryogenesis is the focus of all applied research and it is now considered as the gateway to many more technologies. Plant propagation by somatic embryogenesis not only helps to obtain a large number of plants year round, but also can act as a powerful tool for further transgenic research. In this review we are reporting complete plant production from excised immature zygotic embryos. This method can serve as a support system to conventional and modern agriculture for the genetic improvement of the crop.

INTRODUCTION

Jatropha is a genus of approximately 175 succulents, shrubs and trees (some are deciduous, like *Jatropha curcas* L.), from the family Euphorbiaceae. Though native to North America, Africa and the Caribbean the species is almost pantropical now, widely planted as a biofuel as well as a medicinal plant which soon tends to establish itself.[1,2] *Jatropha* was spread as a valuable hedge plant to Africa and Asia & *INDIA* by Portuguese traders. [1, 2] Many developing countries cannot afford to use edible oils as an energy source because they are already in short supply. India is one of them. Thus, non-edible oils from underresearched plants such as *Jatropha, Pongamia, Neem, Kusum*, and *Pilu* are being advocated. *Jatropha curcas* (ratanjyot) and P*ongamia pinnata* (karanja) could be used to supplement traditional, highly polluting fuels and provide employment to landless and marginal people [3]. Currently, only 10% of the vegetable oils produced are used in nonfood applications such as lubricants, hydraulic oil, bio-fuel, or oleochemicals for coatings, plasticizers, soaps and detergents [4]. Of the above *Jatropha curcas is* considered most potential source as nonedible biodiesel producing plant in our country because it can be grown on almost any soil type. Intercropping with other economically important crops like *Gloriosa, Piper, Karela, Safedmusli, Sarpagandha, Ashwagandha, Gingiber, Curcuma,*etc. may be possible as an additional source of revenue.

Common names : Jatropha, physic nut, Barbados nut, purging nut, pig nut, fig nut, and it is sometimes referred to as the biodiesel or diesel tree [1, 2].

ECOLOGY

Ranging from tropical very dry to moist through subtropical thorn to wet forest life zones, physic nut is reported to tolerate annual precipitation of 4.8 to 23.8 dm (mean of 60 cases = 14.3) and annual temperature of 18.0 to 28.5°C (mean of 45 cases = 25.2). It can grow at saline to alkaline soils, in arid to semi-arid conditions, low slopes of hilly areas, degraded and abused soils [5, 6].

BOTANICAL DESCRIPTION

Shrub or tree to 6 m in height, with spreading branches and stubby twigs, with a milky or yellowish rufescent exudate. Leaves deciduous, alternate but apically crowded, ovate, acute to acuminate, basally cordate, 3 to 5-lobed in outline, 6.0 – 40.0 cm long, 6.0 – 35.0 cm broad, the petioles 2.5 – 7.5 cm long. Flowers several to many in greenish cymes, yellowish, bell-shaped; sepals 5, broadly deltoid. Male flowers many with 10 stamens, 5 united at the base only, 5 united into a column.Female flowers borne singly, with elliptic 3-celled, triovulate ovary with 3 spreading bifurcate stigmata. Capsules, 2.5 – 4.0 cm long, finally drying and splitting into 3 valves, all or two of which commonly have an oblong black seed [1, 2 and 7]. Chromosomes of the genus studied are of very small size. Chromosomes of the microsporocytes of *Jatropha* were mostly paired as bivalents at first metaphase and separate to 11:11 at first anaphase. *J. curcas* have chromosome numbers of 2n=22 and a base number of x=11. [8]

SALIENT USEFUL FEATURES

The non-edible *Jatropha oil* is used for making candles and soap, in adulteration of olive oil, making turkey red oil .The trees produce 1600 liters of oil per hectare. Estimates of the oil content in seeds range from 35-40% oil and the kernels 55-60% [9].The *seeds* can be used as a remedy for constipation; wounds can be dressed with the sap. Nuts can be strung on grass and burned like candlenuts. [1, 10]The seed as well as the fruit is used as a contraceptive. The *cakes* remaining after the oil is pressed out, can be used for cooking, for fertilizing, and sometimes even as animal fodder, while the *Seed husks* can be used to fuel generators. Extracts from this species have also been shown to have anti-tumor activity [2, 6 and 7]. Boiled *Leaves* acts as a remedy for malaria and fever. "The young leaves may be safely eaten, steamed or stewed." They are favored for cooking with goat meat, said to counteract the peculiar smell. In India, pounded leaves are applied near horses' eyes to repel flies. [2, 6 and 7].Mexicans grows the shrub as a host for the lac insect. Ashes of the *burned root* are used as a salt substitute [10]. It has strong *molluscicidal* activity and has been listed as for *homicide, piscicide, and raticide* as well [2, 6 and 11]. The *latex* was strongly inhibitory to watermelon mosaic virus. *Bark* used as a fish poison. Dark blue dye and wax can be produced from the bark of the plant. Sap stains linen and can be used for marking. [2, 6 and 11].

FOLK MEDICINE

The extracts is used in folk remedies for cancer and is reported to be abortifacient, anodyne, antiseptic, cicatrizant, depurative, diuretic, emetic, hemostat, lactagogue, narcotic, purgative, rubefacient, styptic, vermifuge, and vulnerary. Physic nut is a folk remedy for alopecia, anasorca, ascites, burns, carbuncles, convulsions, cough, dermatitis, diarrhea, dropsy, dysentery, dyspepsia, eczema, erysipelas, fever, gonorrhea, hernia, incontinence, inflammation, jaundice, neuralgia, paralysis, parturition, pleurisy, pneumonia, rash, rheumatism, scabies, sciatica, sores, stomachache, syphilis, tetanus, thrush, tumors, ulcers, uterosis, whitlows, yaws, and yellow fever [7,12]. Latex applied topically to bee and wasp stings [6,13]. *Mauritians* massage ascitic limbs with the oil. *Cameroon* natives apply the leaf decoction in arthritis [6, 7, and 12]. *Colombians* drink the leaf decoction for venereal disease. *Bahamans* drink the decoction for heartburn. *Costa Ricans* poultice leaves onto erysipelas and splenosis. *Guatemalans* place heated leaves on the breast as a lactagogue. *Cubans* apply the latex to toothache. *Colombians and Costa Ricans* apply the latex to burns, hemorrhoids, ringworm, and ulcers. *Barbadians* use the leaf tea for marasmus, *Panamanians* for jaundice. *Venezuelans* take the root decoction for dysentery [6, 7, and 12]. Seeds are used also for dropsy, gout, paralysis, and skin ailments [7]. Leaves are regarded as antiparasitic, applied to scabies; rubefacient for paralysis, rheumatism; also applied to hard tumors. Latex used to dress sores and ulcers and inflamed tongues [6, 7]. Seed is viewed as aperient; the seed oil emetic, laxative, purgative, for skin ailments. Root is used in decoction as a mouthwash for bleeding gums and toothache. Otherwise used for eczema, ringworm, and scabies [6, 13]. Homeopathically used for cold sweats, colic, collapse, cramps, cyanosis, diarrhea and leg cramps [12, 13].

CHEMISTRY

Per 100 g, the seed is reported to contain 6.6 g H_2O, 18.2 g protein, 38.0 g fat, 33.5 g total carbohydrate, 15.5 g fiber, and 4.5 g ash [7, 14]. The seed oil belongs to the oleic or linoleic acid group and the major fatty acid composition is oleic (C18:1) 38.54%, linoleic (C18:2) 33.08%, palmitic (C16:0) 16.02% and stearic (C18:0) 10.21% (Table 1). Leaves, which show antileukemic activity, contain a-amyrin, b-sitosterol, stigmasterol, and campesterol, 7-keto-b-sitosterol, stigmast-5-ene-3-b, 7-a-diol, and stigmast-5-ene-3 b, 7-b-diol. Leaves contain isovitexin and vitexin. From the drug (nut) saccharose, raffinose, stachyose, glucose, fructose, galactose, protein, and an oil, largely of oleic- and linoleic-acids, curcasin, arachidic-, linoleic-,myristic-oleic-, palmitic-, and stearic-acids are also reported. Four *antitumor* compounds, including jatropham and jatrophone, are also reported [6, 7, 10, and 11].

A better understanding of the molecular biology of seed-oil accumulation is emerging from studies in other oil-seed crops, particularly members of the genus *Brassica*. The mustard oil is primarily composed of very-long-chain-unsaturated fatty acids (VLCUFA) that account for between 40 and 50% of the oil profile (Table 1) [3]. Based on the recommendation of FAO/WHO for dietary fats in human nutrition, the fatty acid composition of improved edible oil should have a high ratio of MUFA/ SFA, a significant proportion of two essential PUFAs, i.e. C18:2 (x-6) and C18:3 (x-3), with a desirable ratio between 5:1 and 10:1 [3].Thus, mustard oil plant growing in many countries does not have the ideal composition of fatty acids recommended for human energy and nutrition. In particular, erucic acid is nutritionally undesirable [3] and the high-eruate mustard oil is detrimental to mammalian health [3]. Seed oils, which are composed mainly of triacylglycerols (TAG), are an important source of fatty acids for human nutrition and hydrocarbon chains for industrial products [3]. Developing and understanding of the control of TAG synthesis in seeds is an important challenge if yields are to be increased. *Pongamia* oil has higher quantities of unsaponifiables than *Jatropha* oil, while the acid value variation is similar for both. Presence of high unsaponifiable matter content in *Pongamia* hinders its processing for biodiesel production [4]. The fatty acid composition of *Jatropha* seed oil is richer in comparison to the most used edible oil in the context of mustard (Table 1). The absence of the erucic and the behenic acid from the *Jatropha* oil makes it a more suitable proponent for future metabolic engineering studies to bridge the gap of edibility and non-edibility. The following table provides information about the fatty acid composition of three oil yielding crop.

Table 1. Fatty Acid Composition of *Brassica sp*, *Jatropha* and *Pongamia* seed oils

Fatty acids	*Brassica sp* (edible) (%)	*Jatropha* (%)	*Pongamia* (%)
C 14:0	--	1.72	--
C 16:0	1-3	16.02	5.28
C 18:0	0.4-3.5	10.21	7.92
C 20:0	0.5-2.4	0.43	2.91
C 22:0		--	4.89
C 24:0		--	1.85
C 18:1	12-24	38.54	53.14
C 18:2	12-16	33.08	13.47
C 18:3	7-10		--
C 20:1		--	10.54
*Erucic acid	40-55	--	--
*Behenic acid	0.6-2.1	--	4.2-5.3

* Nutritionally undesirable.

TOXICITY

The poisonous property of the plant is mainly due to presence of toxalbumin called curcin and cyanic acid.[15, 16] Though all parts of the plant are poisonous, seeds have the highest concentration of the toxin and are highly poisonous [15, 16]. The adverse effects following consumption of seeds result primarily from gastroenteritis as seen in the present study. The clinical symptoms include vomiting, diarrhoea, abdominal pain, and burning sensation in the throat. Depression and circulatory collapse have also been reported and are said to be common in children.[15, 16] Human deaths by this plant have not been reported so far though animal deaths can occur.[15, 16] The toxic dose is not known. In some instances consumption of as few as 3 seeds has produced toxic symptoms; while in others, as many as 50 seeds produced relatively mild symptoms. Though it is commonly believed that roasting detoxifies the seeds, catastrophes have been reported after eating roasted seeds [15, 16].

Table 2. *Jatropha* Biodiesel properties compared with petroleum diesel and EU standards

Property	Units	*Jatropha* biodiesel	E.U.standards for biodiesel
Density@20C	g/ml	0.879	0.86-0.90
Flash point	^{0}C	191	>101
Cetane number	-	57-62	>51
Kinematic Viscosity(at 40^0C)	cST	4.20	3.5-5
Net calorific value	MJ/Kg	39.5	Undefined
Iodine number	-	95-106	>120
Sulphated ash	%	0.014	>0.02
Carbon residue	%	0.025	>0.3

THE ADVANTAGES OF BIO DIESEL

Bio Diesel from *Jatropha curcas* is the most valuable form of renewable energy that can be used directly in any existing, unmodified diesel engine. Bio fuel from *Jatropha curcas* is (i) environment friendly and ideal for heavily polluted cities (ii) is as biodegradable as salt (iii) produces 80% less carbon dioxide and 100% less sulfur dioxide emissions, non-flammable with very high flash points and provides a 90% reduction in cancer risks (iii) can be used alone or mixed in any ratio with mineral oil diesel fuel. The preferred ratio of mixture ranges between 5 and 20% (B5 - B20) (iv) extends the life of diesel engines (v) is cheaper then mineral oil diesel (vi) and is conserving natural resources and greening of wasteland, drought proofing, energy security for the country and promotion of organic farming [9].Table 2 gives a comprehensive parameter of *jatropha* biodiesel with EU standards for biodiesel[17].

THE ADVANTAGES AND DISADVANTAGES OF CONVENTIONAL CULTIVATION

Jatropha is a seed bearing plant and it can produce 1-2 kg of seed per plant / year when the plant is 2-3 years old. The production amount may increase with the increase with the age of the plant. The edaphic factors also play a role in the rate of seed production. Seed is the production site of biodiesel. However the main disadvantage of *in vivo* cultivation of *Jatropha curcas*, is that it produces an unsure yield of nuts only during a short period of time (once or twice a year) and may not produce optimal yields for several years, which proves to be unsustainable. The plant is cross-pollinated and thus the seeds are of unknown genetic potential. Trees propagated by cuttings show a lower longevity and possess a lower drought and disease resistance than those propagated by seeds [1]. Trees produced from cuttings do not produce true taproots (hence less drought tolerant), rather they produce pseudo-taproots that may penetrate only 1/2 to 2/3rds the depth of the soil as taproots produced on trees grown seed[1]. Clonal propagation by cuttings or other methods that are known to produce high yielding varieties of seeds most often results in cytoplasmic uniformity, making high-density *Jatropha* plantings more susceptible to insect and disease infestation. Several research groups have been working on the improvement of *Jatropha* through conventional selection or breeding programs but the biomass productivity was rather low compared to other oil crops. Therefore if investigation of its genetic diversity and its yield potential can be covered by adequate scientific research, the problem of low yield can be overcome. Application of biotechnology will expedite this process. Importantly, biomass and oil yield can be improved by exploiting genetic engineering focusing on the optimization of enzymes involved in oil biosynthesis and biomass production. However, a prerequisite for the implementation of genetic manipulation of *Jatropha* is the establishment of a robust and efficient transformation procedure. The chapter will focus on recent developments in establishing a transformation procedure for *Jatropha* and present strategies to improve oil quality and yield by genetic engineering.

ECONOMICS OF BIODIESEL FROM *JATROPHA CURCAS*

The by products of biodiesel from *Jatropha* seed are the oil cake and glycerol which have good commercial value. These bye-products shall reduce the cost of biodiesel depending upon the price which these products can fetch. The cost components of biodiesel are the price of seed, seed collection and oil extraction, oil trans-esterification, transport of seed and oil. The cost of biodiesel produced by trans-esterification of oil obtained from *Jatropha curcas* seeds will be very close to the cost of seed required to produce the quantity of biodiesel as the cost of extraction of oil and its processing in to biodiesel is recoverable to a great extent from the income of oil cake and glycerol which are bye-products.The cost of biodiesel is largely dependent on the choice of feedstock and the size of the production facility. Taking these elements into account, the price of biodiesel has been worked out assuming cost of seed as Rs. 5 per kg, 3.28 kg of seed giving one litre of oil and varying prices of by-products. The cost of biodiesel varies between Rs. 16.59 to 14.98 per litre if the price of glycerol varies between Rs 60 and 40 per Kg in the Indian context. [9].

BIODIESEL SCENARIO IN INDIA

Energy, however, remains the mainstay for all civilized world. The priorities may lay in cost economics, environment friendliness, import substitution, or self-sufficiency as strategic objective. Energy is one of the priority areas for the Nation. Indian scenario is unique and different from other developed or developing countries. It has vast areas that are wastelands and are not being utilized for cultivation since these are unfertile, dry, sodic, saline, or alkaline. India has one of the fastest growing economies in the world and fuel consumption is rising with an average of around 5% per year [18]. The rapid growth of India's transport sector is increasing the dependence of India's oil imports. Half of India's oil needs are related with the transportation sector. India imports more than 70 per cent of its crude petroleum needs [18]. *Jatropha*, an oil-bearing plant, which grows in wastelands and requires little care, has long been touted as a potential answer to India's energy problems. Yet, the idea of exploring the commercial possibilities of this hardy plant has hereby remained confined to seminars and field reports in this country. With its vast wastelands, and the global community increasingly rooting for alternate fuels, India could well become the global sourcing hub for both feedstock and processed bio-diesel. As India is deficient in edible oils, non-edible oil is the main choice for producing biodiesel. According to the Indian government policy and Indian technology efforts, some developmental works have been carried out with regards to the production of transesterified non edible oil and its use in biodiesel. Generally a blend of 5% to 20% is used in India (B5 to B20). Indian Oil Corporation has taken up Research and development work to establish the parameters of the production of transesterified *Jatropha* vegetable oil and use of bio diesel. Research is also being carried out for marginally altering the engine parameters to suit the Indian *Jatropha* seeds and to minimize the cost of transesterification. Cultivation for oil in degraded, waste, abandoned and abused lands will provide sustainability, employment generation, and much needed oil to replace fossil fuels. The development of the nation is intricately interwoven with sustainability in energy. In every step of cultivation and processing it has good potential for employment generation, and particularly rural women may see some light to increase their family income. As per planning commissions report (2005), 2.5 million manpower will be required for plantation, 0.75 million for maintenance and 0.1 million for operation in biodiesel units within 2006-07 and it will increase substantially within 2011-12.

BIOTECHNOLOGICAL IMPROVEMENT OF *JATROPHA CURCAS*

The chapter reviews some of the highlights of modern plant biotechnology and discusses the potential applications of biotechnology in the improvement of *J.curcas*. Plant biotechnological approches are considered to be the most appropriate viable alternative.It can facilitate multiple durable resistances to pests and diseases and likewise, transgenes or marker-assisted selection may assist in the development of high yielding crops, which will be needed to meet the demand and need. The genetic base of crop production can be preserved and widen by an integration of biotechnology tools in conventional breeding. Similarly targeting specific genotypes to particular cropping systems may be facilitated by understanding specific gene-by-environment interaction(s) with the aid of molecular research. High quality crops with added-value may be obtained through multidisciplinary co-operation among plant breeders, biotechnologists, and other plant scientists. The chapter restricts its discussion to the major techniques employed in plant regeneration, such as micropropagation, somatic embryogenesis, zygotic embryo culture and genetic engineering.

1. Micropropagation Studies in *Jatropha*

1.1. Regeneration of Plants

The non availability of sufficient cuttings is a major impediment in promoting clonal planting on a commercial scale. Non-availability of superior clones/varieties, shortage of cuttings, low clonal multiplication rate, gaps in knowledge of clonal technology, inadequate evidence about success of clonal plantations, higher cost of clonal plantation, etc. are the major factors that limit large-scale application of clonal technology in India. By using proper explants and proper plant growth regulators it is possible to induce organogenesis or embryogenesis directly from explants or via callus.

Figure 1.
a.   Axillary shoot bud proliferation from nodal explants of *Jatropha curcas.*
b.   Cultures produced innumerable shoot buds per nodal explant after 4–6 weeks.
c.   The regenerated microshoots increased in number as well as in length.
d.   Embryogenic calli obtained from leaf explants of *Jatropha curcas.*
e.   Different stages of germinated somatic embryos.
f.   A complete somatic embryo derived plantlet.
g.   2-3 years old clonally propagated plant in field condition.
h.   2-3 years old somatic embryo derived plant in field.
i.   2-3 years old somatic embryo derived plant in flowering condition.

Clonal Propagation Using Nodal Explants

In vitro clonal propagation of *Jatropha curcas* L. was achieved employing nodal explants (Fig 1a). Axillary shoot bud proliferation was best initiated on Murashige and Skoog's (MS) basal medium supplemented with 22.2 μM N6-benzyladenine (BA) and 55.6 μM adenine sulphate,(Fig 1b) in which cultures produced 6.2 ± 0.56 shoots per nodal explant with 2.0 ± 0.18 cm average lengths after 4–6 weeks (Table 3). The rate of shoot multiplication was significantly enhanced after transfer to MS basal medium supplemented with 2.3 μM 6-furfuryl amino purine (Kn), 0.5 μM indole-3-butyric acid (IBA) and 27.8 μM adenine sulphate for 4 weeks. Both shoot number (30.8 ± 5.48) and average shoot length (4.8 ± 0.43 cm) were found to increase significantly (Table 3) (Fig 1c). Tissue culture studies were undertaken in different species of *Jatropha*. Morphogenesis from endosperm tissues has been reported in *J. panduraefolia* [19, 20, and 21]. High frequency regeneration from various explants of *J. integerrima* has been reported [22]. Using different explants, plant regeneration protocols have also been described in *J. curcas* [23, 24, and 25], but multiplication rate was low for field applications. Moreover, no lab-to-land transfer protocol of *J. curcas* using nodal meristems is available. Nodal meristems are an important source tissue of micropropagation and plants raised from these are comparatively more resistant to genetic variation [26]. Keeping in mind the economical importance of *J. curcas*, critical analysis of the earlier protocols necessitated formulating a well-documented, reproducible, *in vitro* micropropagation protocol. Initiation of cultures from the nodal explants did not pose a major problem. During initiation, the explants did not show any leaching or browning of tissues. MS basal medium was the most effective for *in vitro* shoot multiplication from nodal explants. The nodal explants cultured on MS basal medium supplemented with 55.6 μM adenine sulphate and different cytokinins showed varied response (Table 3). However nodes cultured on half strength MS basal medium showed no visible signs of tissue differentiation. This was possibly due to a greater demand of nitrogen and potassium-containing compounds, which induce greater amount of new proteins [27]. These components are lower in half strength MS basal medium compared to full strength MS basal medium. Micropropagation of *J.curcas* has been reported by various authors [23, 24 and 25, 28, 29], using different tissues except for nodal explants from the field grown plants, but in all of the cases the multiplication rate was low for application. An efficient micropropagation protocol using nodal explants from field grown plants has been achieved in our laboratory [30]. *Jatropha* belongs to the family Euphorbiaceae and our observation confirms earlier reports [31, 32] that among the cytokinins BA plays an important role in initiation of shoot-bud proliferation in many members of the Euphorbiaceae family. However, it has been observed in *Jatropha* that it requires higher concentration of only one type of cytokinin (BA) for induction phase and favours lower concentration of another type of cytokinin (Kn) along with other additives for escalation and proliferation of shoot cultures. A recent report [23] indicated similar effect in *J. curcas* with BA and TDZ, but the authors obtained lesser number of shoots (2.0 ± 0.8 to 12.3 ± 1.7) with comparatively higher amount of cytokinin. The present study describes a well-documented and reliable micropropagation protocol of *J. curcas* from nodal meristems with much higher rate of multiplication. This protocol can be used as a basic tool to commercialize cultivation of the biodiesel plant.

Table 3. Effect of plant growth regulators on shoot multiplication from nodal explants of *J. curcas*

*Growth regulators (µM)			Number of	Length of
BA	IBA	Kinetin	shoots / explant	shoots
2.2	0	0	1.0 ± 0.01^a	0.8 ± 0.02^a
4.4	0	0	2.0 ± 0.13^b	0.8 ± 0.05^a
11.1	0	0	3.0 ± 0.25^a	1.0 ± 0.05^a
22.2	0	0	6.5 ± 0.60^c	2.2 ± 0.19^b
2.2	0	2.3	1.0 ± 0.02^a	0.9 ± 0.06^a
4.4	0	4.6	1.2 ± 0.05^a	0.8 ± 0.05^a
11.1	0	11.6	1.0 ± 0.03^a	0.9 ± 0.08^a
22.2	0	23.2	1.0 ± 0.01^a	0.8 ± 0.04^a
2.2	0.5	0	1.2 ± 0.08^a	1.0 ± 0.05^a
4.4	1.0	0	1.5 ± 0.04^a	0.8 ± 0.06^a
11.1	2.5	0	1.3 ± 0.03^a	0.9 ± 0.03^a
22.2	5.0	0	1.0 ± 0.02	0.9 ± 0.01^a
0	0.5	2.3	30.8 ± 5.48^g	4.8 ± 0.43^d
0	1.0	4.6	14.6 ± 2.93^f	3.9 ± 0.48^c
0	2.5	11.6	11.2 ± 2.73^e	3.6 ± 0.32^c
0	5.0	23.2	9.8 ± 1.86^d	3.0 ± 0.29^c
0	0	2.3	2.4 ± 0.25^a	0.9 ± 0.04^a
0	0	4.6	2.2 ± 0.22^a	0.9 ± 0.06^a
0	0	11.6	2.6 ± 0.28^a	0.9 ± 0.05^a
0	0	23.2	$3.7 \pm .36^b$	1.0 ± 0.08^a

* MS basal medium +3% sucrose + adenine sulphate (27.8 µM) along with different PGR.
Data was recorded after 3-4 weeks of culture. Each treatment was repeated three times and each replicate consisted of 3-5 explants. Means having different letters in superscript are significantly different from each other (P≤ 0.05) according to Duncan's multiple range test.[30]

1.2. Somatic Embryogenesis

Somatic embryogenesis involves *in vitro* formation of embryos from somatic tissues cultured on media supplemented with various hormones. These embryos develop from single cells and pass through globular, heart-shaped, torpedo-shaped and cotyledonary embryo stages to give rise to complete plantlets. The whole process may proceed on a single medium or may require medium alterations specific for a developmental stage. A single cell can give rise to numerous propagules. Since a plantlet is derived from a single cell, therefore, there is higher possibility of obtaining true transgenics. Somatic embryogenesis, a powerful tool of plant biotechnology for faster and quality plant production has been successfully applied to regenerate plants in *Jatropha curcas* for the first time. Embryogenic calli were obtained from leaf explants on MS basal medium [33] supplemented with only 9.3 µM Kn (Fig 1d). Induction of globular somatic embryos from 58% of the cultures was achieved on MS medium with different concentrations of 2.3–4.6 µM Kn and 0.5–4.9 µM IBA; 2.3 µM Kn and 1.0 µM IBA proved to be the most effective combination for somatic embryo induction in *Jatropha curcas* (Table 4) (Fig 1e). Our study required a minimum time of 4–6 weeks for conversion of globular somatic embryos to germinated plantlets (Fig 1e). Addition of 13.6 µM adenine sulphate along with 2.32 µM Kn and 1.0 µM IBA led to the significantly (P < 0.05) highest mean number of mature somatic embryos (24.0 ± 6.8) after 4 weeks of culture, which served the best concentration and combination of PGRs (Table 4). Mature somatic embryos were converted to plantlets on half strength MS basal medium with 90% survival rate in the field condition (Fig 1f). The whole process required 12–16 weeks of culture for completion of all steps of plant regeneration. An interesting observation was the induction of a low frequency of secondary embryogenesis along the shoot and root poles of well-developed primary embryos grown in the maturation medium [34]. The synergistic combination of reduced Kn, IBA and adenine sulphate not only promoted the growth of shoot and root, but also secondary embryo formation. In this study, both proliferation and germination of secondary embryos are successfully reported for the first time [34]. This protocol of somatic embryogenesis in *Jatropha curcas* may be an ideal system for future transgenic research and metabolic engineering studies.

Table 4. Number of somatic embryos, germination and establishment of somatic plantlets from 4 week old embryogenic cultures of *Jatropha curcas*

MS basal medium +3% sucrose (w/v)+ concentration of growth regulator +13.6µM AD.		Percentage of embryogenesis (%)	Number of somatic embryos ± S.E.	Number of somatic embryos germinated ±S.E.	Number of somatic plantlets recovered ±S.E.
Kinetin (µM)	IBA (µM)				
2.32	0.5	72	40.0±10.6cd	18.0±4.5b	2.0±0.56b
	1.0	80	58.5±12.7d	24.0±6.8b	5.0±0.87c
	2.5	67	32.6±8.9c	15.5±4.6c	2.0±0.22b
	4.9	58	37.8±9.2c	10.0±3.0b	2.0±0.35b
3.5	0.5	67	31.7±8.3b	12.5±3.2b	1.5±0.24ab
	1.0	53	25.5±7.4b	10.0±3.4a	1.0±0.05a
	2.5	58	22.8±7.2b	16.0±4.9a	1.0±0.08a
	4.9	50	23.0±7.9b	8.5±2.3a	1.0±0.04a
4.6	0.5	60	16.0±4.3a	0	0
	1.0	57	11.5±3.8a	0	0
	2.5	48	8.0±2.7a	0	0
	4.9	42	5.6±1.8a	0	0

Data was recorded after 6-12 weeks. Each treatment was replicated three times, and each replicate consisted of 20 explants. Means having different letters in superscript are significantly different from each other (P< 0.05) according to Duncan's multiple range test.[34]

1.3. Embryo Excision and Culture

The culture of an embryo isolated from the seed and ovules of higher plants in special medium is defined as embryo culture. Embryo culture is used for various purposes. These include basic studies on embryology, breaking the seed dormancy, testing the vitality of seeds, production of rare species, rescue of embryos and production of haploid plants. Embryogenesis *in vitro* represents a powerful tool to manipulate plant development [35]. For culture of embryos; seeds/fruits are surface sterilized before embryo excision. The chapter deals with the effects of medium and sterilization on the immature embryo culture of *Jatropha curcas*. In our laboratory, the green fruits were washed thoroughly in tap water and with 5% Teepol for 30 minutes and washed repeatedly. Then the whole fruits were treated with 0.1 %(w/v) bavistine solution for 15 minutes, washed thoroughly and followed by a 10 minutes treatment in 0.1% $HgCl_2$ for immature seeds (Fig 2a). Then it was washed thoroughly about 4 to 5 times in sterile distilled water so that no traces of the sterilant remains. Before the embryos are removed the sterilized seeds were further dipped in 95% ethanol and flamed to provide maximum aseptic condition for the embryo excision procedure. The most important aspect of embryo culture is determining a culture medium that will provide the regular growth of embryos. In our studies, the effects of different media on the development of embryos have been investigated and usually MS (Murashige & Skoog) [33], WPM (Woody Plant Medium) [36] and B5 (Gamborg's medium) [37] with 3% sucrose have been used for the embryo culture. A sucrose concentration lower or higher than 3% resulted in lower germination or promoted callus formation. Among the different media used the highest plantlet development from excised embryos was obtained in MS (95.1%), (Fig 2d & 2e) followed by ½ MS (85.3%), B5 (71.5%) and WPM (34%).The woody plant medium (WPM) was not at all suitable for embryo development and thus was discarded. The embryogenic tissue was scooped out of the perisperm (Fig 2b). On the 2nd day of the culturing process in MS medium the embryo germinated with the emergence of the green cotyledonary leaves (Fig 2c). The combination of half strength MS with very low doses of cytokinin also gave similar result. About 95.1% germination rate was revealed within a week. The length of growing plants, root number and the length of the roots of the 2months old hardened plantlets (Fig 2g) were measured and evaluated statistically (Table 5). The germination of embryos and development of plantlets from embryos were evaluated based on media and also sucrose concentration. Reduced concentration of sucrose (1.5% from initial 3%) plays a vital role in the overall development of the plantlets. Normally both root and shoot developments were observed in plantlet development from embryos (Fig 2e).Fig 2f shows a complete rooted plant from zygotic embryo excision. However in some embryos, only shoot development without any root development was observed. An interesting observation was the induction of secondary organogenesis on the cotyledonary leaves when subjected to a PGR combination of 4.6 µM Kn + 1.0 µM IBA + 74.0 µM adenine sulphate (Fig 2i). The zygotic embryo culture technique has been adopted in case of *Jatropha curcas* for the first time to rescue embryos from seed one of which is seen to usually abort and to overcome dormancy of recalcitrant seeds and to reduce the breeding cycle of the plant.

Figure 2.

a   Immature seeds of *Jatropha curcas.*

b. Embryogenic tissue from excised embryo.

c. Germinated zygotic embryo of *J.curcas.*

d. Developing excised zygotic embryo in full strength MS medium.

e. Developed excised zygotic embryo with distinct shoot and root poles.

f. 5-6 weeks 1inch plantlet of *J.curcas* with 5 distinct tap roots.

g. Plantlets regenerated from excised embryo established directly to autoclaved garden soil.

h. A 5 months old hardened embryo excised plant in field condition.

i.   Induction of secondary direct organogenesis on the cotyledonary leaves of the excised zygotic embryo of *J.curcas.*

j.   Chromosomal analysis revealed a diploid parental chromosome number of 2n = 22 very small chromosomes.

Table 5. Tabulated data showing average root number root length, total plantlet size and percentage of plantlet germination of the excised embryo plants

Media	Root Number Mean ± S.E	Root length (cm) Mean ± S.E	Size of plantlet (cm) Mean ±S.E	Rate of plant germination (%)
MS	4.2±0.25	5.63±0.10	9.4±0.17	95.1
½ MS	3.9±0.21	4.2±0.21	7.8±0.22	85.3
B5	2.7±0.20	3.2±0.22	5.7±0.21	71.5

Data was recorded after 8-10 weeks. Each treatment was replicated three times, and each replicate consisted of 20 explants. Data evaluated as mean ± S.E.

Table 6. Effect of IBA on initiation of roots in excised shoots of clonally propagated *J. curcas*

IBA (µM)*	Percentage of root induction (%)	Number of roots / shoot	Length of roots (cms ± S.E.)
0	0	0	0
0.5	15	1.0 ± 0.06^a	1.2 ± 0.09^a
1.0	52	5.6 ± 0.42^b	8.7 ± 1.35^b
2.5	0	0	0
5.0	0	0	0
10.0	0	0	0

* MS basal medium + 3% sucrose.

Data was recorded after 2-3 weeks of culture. Each treatment was repeated three times and each replicate consisted of 3-5 shoots. Means having different letters in superscript are significantly different from each other (P≤ 0.05) according to Duncan's multiple range test.

1.4. In vitro Rooting of Shoots

Plants regenerated from nodal explants showed 52% of root induction in MS basal medium supplemented with 1.0 µM IBA in 2–3 weeks (Table 6). The 8–10-week-old shoots (2.0–3.0 cm in length) were cultured on MS basal medium supplemented with 3% (w/v) sucrose and different concentrations of auxins were tested individually. Further elongation of roots with average length of 8.7 ± 1.35 cm. was obtained in unsupplemented MS basal medium for 2–3 weeks (Table 6). A distinct taproot system developed with slender and white secondary roots; this was considered important for hardening and field transfer. Significant increase in root length occurred on transfer to only MS basal medium for another 2–3 weeks.

1.5. Establishment of the Plantlets

After *in vitro* rooting of shoots, the plantlets have to be established in soil and for this various methods have been employed. Prior to transfer to soil, hardening of the plantlets is done. The plantlets from different regeneration lines were subjected to the following hardening procedures. Plants regenerated directly from nodal explants were cultured on MS 1/4th strength without agar for elicitation of root growth. Root portion was then washed with 0.5% bavistine and tap water. The plantlets were then transferred to garden soil, and vermiculite in the ratio 1:1:1.Fig 1g shows 2-3 years old clonally propagated plant in field condition. The somatic embryo-derived plantlets were taken out of the culture vessels, thoroughly washed with tap water, dipped for 1 h in 0.1% (w/v) bavistin (systemic fungicide) and transferred to plastic pots containing a mixture of sand and vermicompost in the ratio 1:1 and covered with polythene bags. The plantlets were irrigated with tap water as and when required. After 4 weeks these plantlets were transferred to bigger pots containing garden soil mixed with organic manure. When the plantlets showed signs of establishment with the appearance of new leaves, the polythene bags were removed gradually for acclimatization to field conditions. Fig 1h shows a 2-3 years old somatic embryo derived plant in field. Plantlets regenerated from excised embryo were established directly to autoclaved garden soil (Fig 2g).Fig 2h shows 5 months old hardened embryo excised plant. On the basis of all the information obtained by these studies, it can be said that regenerated plantlets should preferably be of small size (3-4 inches) and roots should be clearly developed for successful establishment. Soilrite, vermiculite, soil mixtures have been used for their establishment. Plantlets can also be transferred to soil or soilrite. Soil should be kept moist with frequent watering; plants may be covered for 7-10 days to maintain humidity. Temperatures should preferably be 20-22^{0}C.After 10-15 days of hardening in pots the plants can be transferred to the field.

1.6. Chromosomal Analysis

Cytological studies from randomly selected root tips of clonally propagated plantlets and of somatic embryos were carried out using saturated PDB solution as pretreating chemical for 4 h followed by overnight fixation in Carnoy's fluid. Materials were stained with 2% aceto-orcein, HCl (9:1) solution [38] Chromosomal analysis revealed a diploid parental chromosome number of 2n = 22 very small chromosomes (Fig 2j). No cytological anomaly was observed in our studies, indicating a genotypic stability. The stained materials were squashed for microscopic observation, and photographs were taken under Zeiss photomicroscope.

1.7. Statistical Analysis

The experiments were set up in a randomized design. Data were analyzed by analysis of variance (ANOVA) to detect significant differences between means [39]. Means differing significantly were compared using Duncan's multiple range test (DMRT) at a 5% probability level. Variability of data has also been expressed as the Mean ± standard error (SE).

1.8. Floral Biology

Male and female flowers are produced on the same plant but at separate sites on the plant. The general ratio of female is to male flowers are 1/5 or even 1/7, rarely 1/3, the female flower being in the centre surrounded by the male flowers. Flowering depends on the climatic conditions and the nature of the plant. Short day length causes an increased formation of male flowers. In regions with dry seasons flowering is associated with the rainy season. The plants are well adapted to drought, shedding of leaves. The pollination is by insects. The fruits generally trilocular capsule containing three seeds, are produced in bunches. Ripe fruits changes colour from green to yellow. The seeds are ellipsoid and black 1.5-2.5 cm in length and 1.0 cm in breadth.[1,2,6].

1.9. Flowering Time and a Step towards Reduction of the Breeding Cycle of the In Vitro Grown Plants

Seed setting requires 24 to 36 months time in *Jatropha curcas*. The mother or donor plant in our studies began flowering after 24 months. On the other hand somatic embryo derived plant growing in the field flowered after 19 to 20 months of transfer in the field (Fig 1i) which is about 4-5 months earlier than the donor plant. We are now observing the seed setting pattern in this plant. Early flowering in *in* vitro grown *Jatropha curcas* will be a very important parameter and has not been reported yet. The fruits taken in culture were obtained after 7-8 weeks of flowering. The excised embryos in culture germinated within a week which would have taken atleast 15-18 weeks conventionally (Fig 2c). The following flow chart compares seed germination rate of the excised embryo plant and the mother plant.

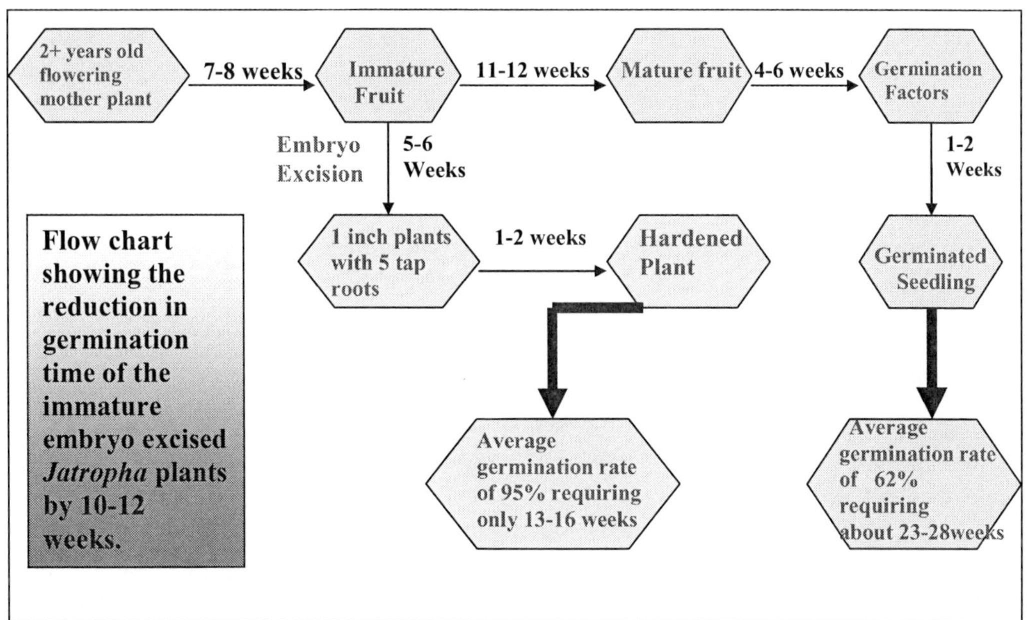

2. Genetic Engineering Studies in *Jatropha Curcas*

Biotechnological studies in *Jatropha curcas* are limited with developing interest in recent years. Genetically modifying *Jatropha* could be a short cut to higher yields in less time. Through our studies it was revealed that the regeneration response is a complex phenomenon which is governed by a number of factors. In recent years, with the advent of recombinant DNA technology and PCR, molecular markers are being used for a variety of studies. Genetic analysis in species with a view to identify quantitative trait loci (QTLs) for useful characters inherited in a multigenic fashion is now possible even without prior information of allelic differences because of the development of molecular techniques such as RAPD and AFLP, which directly detect variation at the DNA level rather than at the phenotypic level[4]. To begin with, an elite mapping population with the desirable characteristics that breeds true would have to be established in *Jatropha*. Information from the elite mapping population data could then be used to rapidly screen germplasm of *Jatropha* to identify DNA markers or major QTLs associated with high yield, and the same could be used in marker assisted selection (MAS) breeding strategies to jumpstart genetic improvement of *Jatropha* for yield characteristics [4]. The molecular markers include Restriction Fragment Length Polymorphism (RFLP's), Random Amplified Polymorphic DNA's (RAPD's), Amplified Fragment Length Polymorphism (AFLP's), Simple Sequence Repeats (SSR's) or Sequence Tagged Microsatellite Sites (STMS), Sequence Tagged Sites (STS), DNA Amplification Fingerprinting (DAF) and Microsatellite Primed-PCR (MP-PCR). The molecular markers most recently are applied for testing and confirmation of clonal fidelity of the *in vitro* grown plants and ex situ conservation and characterization of the plant genetic resources. For commercial industrialization of the micropropagation techniques the foremost concern is the maintenance of the true-to-type nature of the micropropagated plants *vis-a-vis* explant source. This can only be achieved by developing micropropagation protocols based upon axillary branching and somatic embryogenesis which has already been discussed in this chapter. There is virtually no information with regard to the number of introductions and the genetic diversity of *Jatropha curcas* populations grown in India. Several researches have attempted to define the origin of *J.curcas*, but the source remains controversial [1,40].In Euphorbiaceae, molecular markers such as RAPD, RFLP and SSRs have been employed for determining the extent of genetic diversity in elite rubber clones [41,42].In *Jatropha* isozyme markers were used to determine the genetic relatedness of the genus *Jatropha* and *Ricinus* [43].RAPD markers were employed to confirm hybridity of inter-specific hybrids and to determine the similarity index between one accession each of the toxic Indian and non-toxic Mexican genotypes [44]. Genetic diversity in *Jatropha curcas* germplasm from India along with a non-toxic Mexican genotypes was evaluated using specifically the RAPD and ISSR primers [45] showing a narrow genetic base. However in a recent report a much higher rate of polymorphisms with 26 RAPD primers was detected [46]. Therefore it can be rightly said that species has not been improved for productivity and most of the projects relied on natural genotype occurring as unadapted populations. Furthermore detailed molecular characterizations of the tissue culture grown plants are not yet available and hence require elucidation and certification. Hence there is an immediate need for systemizing research for widening the genetic base of *J.curcas* through selection of superior genetic stocks, mutagenesis and inter-specific hybridization. The main thrust of genetic improvement of the crop should be on high yield, high oil content, early maturing varieties and reduction of toxicity.

FUTURE RESEARCH NEED

Future research should be directed towards multifarious line of research like evolving efficient transformation protocols involving RNAi technology. Gene pyramiding is the latest trend which can also be applied to *Jatropha curcas* . Metabolic engineering of the plant may show its potential both in basic research and as a tool of modern plant breeding. Designer crops will be able to produce valuable enzymes, proteins, and antibodies. The large-scale production of useful plant material or metabolites in bioreactors shows new possibilities in plant biotechnology. Research and development should aim at gene prospecting for increased oil synthesis, metabolic engineering of oil biosynthesis pathway and molecular markers for marker aided selection. Metabolic engineering is a powerful tool to alter and ameliorate the secondary metabolite composition of crop plants and gain new desired traits. The interplay of a multitude of biosynthetic pathways and the possibility of metabolic cross-talk combined with an incomplete understanding of the regulation of these pathways, explain why metabolic engineering of plant secondary metabolism is still in its infancy and subject to much trial and error. There is a possibility to increase the yield of oil from *Jatropha* seed through transgenesis. The study of development of seed, endosperm and oil bodies of *Jatropha* will be an essential step forward to transgenesis. The information on endosperm and oil development of *Jatropha* seeds will greatly facilitate the study of gene expressions that are involved in biosynthesis of storage lipids. Genetic approaches to investigating the regulation of oil content have so far met with limited success. The fatty acid composition of seed oil varies considerably both between species and within species, with fatty acids varying in both chain length and degrees of desaturation. The target would be to isolate the key genes which are involved in de novo fatty acid/lipid biosynthesis through molecular and genomic approaches. We aim to genetically engineer fatty acids (FA)/lipid biosynthesis in *Jatropha* seeds to increase oil yield and unsaturated FA in storage lipids. Altering the seed oil composition will aid in achieving the desired density, dynamic viscosity, heat capacity & boiling point and heat of combustion. The lower dynamic viscosity and higher heat of combustion is mainly attributed by medium and short chain fatty acids in the seed oil. To develop economically viable oilseed crops with modified fatty acid profiles, there is a requirement to manipulate the activity (or gene expression) of relevant key constituent steps in the synthetic or modification pathways, i.e. to carry out genetic metabolic engineering. This can be achieved either through up- or down-regulation of an introduced recombinant gene (transgenic), deletion of endogenous genes (mutagenesis), or by selection of appropriate combinations of the relevant naturally occurring alleles present in the gene pool. Therefore metabolic engineering of the fatty acid biosynthetic pathway to produce a greater amount of medium chain fatty acids is indeed a feasible approach. *Jatropha* oil is a complex mixture of four fatty acids (oleic, linoleic, palmitic, stearic acids) that have vastly differing melting points, oxidative stabilities, and chemical functionalities. In this regard, genetic engineering approaches can be applied to enrich the content of *Jatropha* oil for a particular fatty acid or class of fatty acids. *Jatropha* oil can be comparable to olive oil in its oleic acid content. Edible oils with high oleic acid content are reported to be beneficial as oleic acid increases the stability of oil and affects fat content of blood favourably, [3]. The most notable point of research will be to produce transgenic *Jatropha* seeds with enhanced rate of oleic or linoleic acid. This can be obtained either by down-regulation or over-expression of a particular gene. Further modification can be done using genetic engineering to down regulate conversion of oleic acid into linoleic acid and omega 3 fatty acid. Thus, new cultivar with desired percentage of oleic acid and linoleic acid and omega 3 fatty acid can be produced. High-oleic oils with elevated oleic acid content are generally considered to be healthier oils. Enormous potential lies if this high – oleic oil variety can be bred with non-toxic species to raise cultivar that can be used for production of both edible oil and biodiesel. Bioengineering methods has been partially possible in one of its family member *Cassava* containing potentially toxic levels of the cyanogenic glycosides [namely linamarin (95%) and lotaustralin (5%)]. Inhibiting linamarin synthesis in leaves and its accumulation in roots accelerated cyanogenesis and food detoxification.[47]

CONCLUSION

This chapter was an attempt to define the scope of employing biotechnological tools for improving the productivity of biofuel oilseed crop *Jatropha curcas* and production of biodiesel. The commercial use of by-products has the advantage of hardly any waste being produced and of optimised financial feasibility. The future line of research targets the monitoring of detoxification of seed beans and oil , utilization of the residue de-oiled cake for fodder and manure , conversion of non-edible to edible oil. The seeds of *Jatropha curcas* L. contain 4% to 40 % oil with a fatty acid pattern similar to that of edible oils. Metabolic engineering and RNAi technology can also be considered for further improvement to achieve the goal. For genetic improvement tissue culture protocols for the rapid propagation and regeneration of selected genotypes of *Jatropha curcas* are highly desirable. The molecular characterization of the tissue culture plants is not yet available and needs immediate action. The major drawbacks lie in the lack of studies on the breeding and commercial processing of *Jatropha curcas* as source of biodiesel. Intercropping with other economically important crop will provide additional revenue. Laboratory evaluation on the agronomic performance of the cultivars including the extractability and quality of biodiesel, macro and micro propagation methods of superior cultivars for commercial production, germplasm management, varietal improvement and development of esterified, non toxic oil and diesel extraction technologies for *Jatropha* seeds is very much essential.

REFERENCES

[1]   Heller, J. (1996). *Physic nut. Jatropha curcas L. Promoting the conservation and use of underutilized and neglected crops.* Institute of Plant Genetics and Crop Plant Research (IPGRI), Gatersleben/International Plant Genetic Resources Institute, Rome.

[2]   Openshaw, K. (2000). A review of *Jatropha curcas*: an oil plant of unfulfilled promise. *Biomass Bioenerg.19*:1-15.

[3]   Ohlrogge, J & Browsw, J. (1995).Lipid biosynthesis. *Plant Cell.7*: 957-970.

[4]   Nambisan, P. (2007).Biotechnological intervention in jatropha for biodiesel production. *Current Science.93 No 10*:1347-1348.

[5]   Francis, G. Edinger, R. Becker, K. (2005). A concept for simultaneous wasteland reclamation, fuel production, and socio-economic development in degraded areas in India: Need, potential and perspectives of *Jatropha* plantations". *Natural Resources Forum 29:* 12–24.

[6]   Duke (www.tropilab.com/jatropha-cur.html) TROPILAB INC's webpage.

[7]   Duke, J.A. (1983) *Jatropha curcas* L. (Euphorbiaceae) Physic nut, Purging nut. *Handbook of Energy Crops*. Unpublished. (Personal correspondence with the author on 04/10/06).

[8]   Soontornchainaksaeng, P. & Jenjittikul, T. (2003). Karyology of *Jatropha* in Thailand. *Thai For. Bull. 31*:105-112.

[9]   http//www.jartropha.org

[10]   Senegal *Jatropha* Oil Project. University of Hohenheim. March 1996 to March 2000. Studies on the utilization of *Jatropha curcas* seed cake as an animal feed. Enterprise Works (formerly Appropriate Technology International).

[11]   Henning, R K. (undated) *Jatropha curcas L. in Africa.* Global Facilitation Unit for Underutilized Species. *Rothkreuz 11,* D-88138 Weissenberg, Gemany (henning@bagani.de)

[12]   Ross, I. A. (1999). Medicinal Plants of the World. *Chemical Constituents, Traditional Uses and Modern Medicinal Uses.*

[13]   Wyeth, P. (2002). An Industry and Market Study of Six Plant Products in Southern Africa, *Jatropha* or Physic Nut. International Programs, Washington State University, funded by the United States Agency for International Development.

[14]   Prajapati, N. D. & Prajapati, T. (2005). (Eds). A Handbook of *Jatropha curcas* Linn. (Physic Nut).

[15] Langdon, K. R. (1977) Physic nut, *Jatropha curcas* Nematology (Botany). *No.3* (www.doacs.stat.fl.us/npi/empp/botany/boteire/nem Botaro 30.htm).

[16] Dhyani, S. K. (2006). (Eds).Insect pests of *Jatropha curcas* L. and the potential for their management.

[17] Francis, G. Edinger, R. Becker, K. (2005). A concept for simultaneous wasteland reclamation, fuel production, and socio-economic development in degraded areas in India: Need, potential and perspectives of *Jatropha* plantations. *Natural Resources Forum. 29:* (2005) 12–24.

[18] Mandal, R. (2005). "Energy – alternate solutions for India's needs: biodiesel", Adviser for the Planning Commission, Government of India.

[19] Srivastava, P.S.(1971). *In vitro* induction of triploid roots and shoots from mature endosperm of *Jatropha panduraefolia. Z.Pflanzenphysiol.. 66*, 93-96.

[20] Johri, B. M. & Srivastava, P. S.(1973). Morphogenesis in endosperm cultures. *Z. Pflanzenphysiol. 70*, 285-304.

[21] Srivastava, P. S.& Johri, B. M.(1974). Morphogenesis in mature endosperm cultures of *Jatropha panduraefolia. Beitr. Biol. Planz.. 50*, 255-268.

[22] Sujatha, M. & Dhingra, M.(1993). Rapid plant regeneration from various explants of *Jatropha integerrima. Plant Cell Tiss. Org. Cult. 35.* 293-296.

[23] Sujatha, M. Makkar, H. P. S. & Becker, K. (2006). Shoot bud proliferation from axillary nodes and leaf sections of non-toxic *Jatropha curcas* L. *Plant Growth Reg., 47*, 83-90.

[24] Qin, W. Wei-Da, L. Yi, L. Shu-Lin, P. Ying, X.U. Lin, T. & Fang, C. (2004). Plant regeneration from epicotyl explants of *Jatropha curcas. J. Plant Physiol. Mol.Biol..30.*475-478.

[25] Sujatha, M. & Mukta, N. (1996). Morphogenesis and plant regeneration from tissue cultures of *Jatropha curcas. Plant Cell Tiss. Org. Cult. 44*, 135-141.

[26] Pierik, R. L. M. (1991) (eds).Commercial aspects of micropropagation. In *Horticulture-New Technologies and Applications* Dordercht, Netherlands.

[27] Guru, S. K. Chandra, R. Khetrapal, S. Raj, A. & Palisetty, R. (1999).Protein pattern in differentiating explants of chick pea (*Cicer arietinum* L.). *J. Plant Physiol. 4.*147-151.

[28] Rajore, S. & Batra, A. (2005). Efficient plant regeneration via shoot tip explant in *Jatropha curcas. J. Plant Biochem. Biotech. 14.* 73-75.

[29] Sardana, J. Batra, A. Ali, D .J.(2000). An expeditious method for regeneration of somatic embryos in *Jatropha curcas* L. *Phytomorphology .50:*239–242

[30] Dutta, M. M. Mukherjee, P. Ghosh, B. & Jha, T. B. (2007). *In vitro* clonal propagation of biodiesel plant (*Jatropha curcas* L.). *Current Science. 93. No 10.* 1438-1442.

[31] Ripley, K. P. & Preece, J. E. (1986). Micropropagation of *Euphorbia lathyrus* L. *Plant Cell Tiss. Org. Cult. 5.* 213-218.

[32] Tiedman, J. & Hawker, J. S. (1982). *In vitro* propagation of latex producing plants. *Ann Bot. 49*, 273-279.

[33] Murashige, T. & Skoog, F. (1962) A revised medium for rapid growth and bioassays with tobacco tissue cultures. *Physiol. Plant. 15,* 473-479.

[34] Jha, T. B. Mukherjee, P. & Dutta, M. M. (2007). Somatic Embryogenesis in *Jatropha curcas* Linn., an important biofuel plant. *Plant Biotechnology Reports. 1; No.3.* 135-140.

[35] Monnier, M.(1990). Zygotic embryo culture. Editor S.S.Bhojwani. Plant Tissue Culture: Applications and Limitations. Chapter 16 pp.366-393.*Elsevier Science Publishers* B.V. The Netherlands.

[36] McCown, B.H. & Lloyd, G. (1981). Woody plant medium (WPM). A mineral nutrient formation for microculture of woody plant species. *Hortscience,16*: 453."

[37] Gamborg O.L. & Shyluk J.P. (1981).Nutrition, media and characteristics of plant cells and tissue cultures.In:Thorpe TA (ed) Plant Tissue Culture methods and applications in agriculture.Academic Press,New York, pp 21-44.

[38] Sharma, A.K. & Sharma, A. (1990). Study of plant chromosomes from tissue culture. *Chromosomal techniques theory and practice,* 3rd edn. Aditya Books. pp 317–338.

[39] Sokal, R. & Rohlf, F.J. (1987). Introduction to Biostatistics, 2nd edn. Freeman WH, New York.

[40] Dehgan,B. & Webster,G.L. (1979). Morphology and Intergeneric relationship of the genus *Jatropha*. *Univ of Calif Publ Bot. 74:* 1-7.

[41] Besse, P. Seguin, M. Lebrun, P. Chevallier, M.H. Nicholas, D. Lanaud, C. (1994). Genetic Diversity among wild and cultivated populations of *Hevea brasiliensis* assessed by nuclear RFLP analysis. *Theor Appl Genet.88*:199-207.

[42] Varghese,Y.A. Knaak, C. Sethuraj,M.R. Ecke,W. (1997). Evaluation of random amplified polymorphic DNA (RAPD) on *Hevea brasiliensis*. *Plant Breed. 116*:47-57.

[43] Sathaiah, V. Reddy, T. P. (1985). Seed protein profiles of castor (*Ricinus communis* L.) and some *Jatropha* species. *Genet Agr, 39*:35-43.

[44] Sujatha, M. Prabakaran, A.J. (2003). New ornamental *Jatropha* hybrids through interspecific hybridization. *Genet Resource Crop Evol.50*:75-82.

[45] Basha, S.D. Sujatha,E. M. (2007). Inter- and Intra –population variability of *Jatropha curcas* (l.) characterized by RAPD and ISSR markers and development of population specific SCAR markers.*Euphytica.156*:375-386.

[46] Ganesh R.S. Parthiban, K.T. Senthil, R.K. Thiruvengadam, V. Paramathma, M. Genetic diversity among *Jatropha* species as revealed by RAPD markers. *Genet Resource Crop Evol.* DOI 10.1007/s10722-007-9285-7.

[47] Sayre, R. (2004).Metabolic engineering of cyanogenic glucoside synthesis and turnover in *Cassava*. Dept Plant Cell & Molec.Biol.Ohio State Univ.USA.

In: New Research on Biofuels
Editors: J. H. Wright and D. A. Evans

ISBN 978-1-60456-828-8
© 2008 Nova Science Publishers, Inc.

Chapter 4

ENVIRONMENTAL ASPECTS OF ENERGY CROPS – ENERGY BALANCE, EMISSIONS, AND CARBON SEQUESTRATION

Volkhard Scholz[1,], Hans Jürgen Hellebrand[2,†] and Martin Strähle[3,‡]*

[1]Leibniz-Institut für Agrartechnik Potsdam-Bornim (ATB), Dept. of Post-Harvest Technology, Max-Eyth-Allee 100, 14469 Potsdam, Germany
[2]Leibniz-Institut für Agrartechnik Potsdam-Bornim (ATB), Dept. of Technology Assessment and Substance Cycles, Max-Eyth-Allee 100, 14469 Potsdam, Germany
[3]Technische Universität Berlin (TUB), Institute of Ecology, Dept. of Soil Science Salzufer 11-12, 10587 Berlin, Germany

ABSTRACT

On an experimental field in eastern Germany 10 haulm-type and woody crop species which are adequate for combustion and gasification have been cultivated under practical conditions on a sandy brown soil for 14 years now. Each crop received 4 different levels of fertilization, from 0 kg N ha^{-1} to 150 kg N ha^{-1} combined with straw and wood ashes. The measuring program includes yields, energy gain and environmentally relevant substances in plants and soil, as well as fertilizer-induced emissions. Measured long-term yields are between 7 and 10 t ha^{-1} y^{-1} for all crops with the exception of topinambur and a special poplar variety. In contrast to the yields of haulm-type crops which decrease along with reduced fertilization, the woody crops such as poplar and willow show an increase. Energy yields average out between 100 and 160 GJ ha^{-1} y^{-1}, while the energy demand for cultivation and harvesting only accounts for 1% to 13% of these yields.

---

[*] Correspo nding author. E-mail: vscholz@atb-potsdam.de
[†] E-mail: jhellebrand@atb-potsdam.de
[‡] E-mail: martin.straehle@web.de

The nitrogen content (N_t) varies between the crops in a range from 0.1% to 3.3% and depends on the fertilization level. With 0.2 to 1.2% N_t, poplars and willows have only half the N_t of whole crop cereals and hemp, and approximately only one third that of cocksfoot grass. Therefore, these species cause less NO_x emissions during combustion. Nitrous oxide (dinitrogen oxide: N_2O) flux measurements, carried out over nine years, show that the mean nitrogen conversion factor, which describes the N_2O greenhouse burden of fertilization, is about 0.8±0.1 % and thus slightly lower than the default value of 1% for N_2O inventories. Therefore, the production of lingo-cellulosic energy crop species considered here will not lose its CO_2 advantage by nitrogen fertilizing as long as fertilizing results in an adequately higher biomass yield.

Soil organic carbon (C_{org}) stocks are likely to be improved by a land-use change from annual to woody crops. Twelve years after establishment of the plantation, a comparison of C_{org} stocks under annual crops with C_{org} under short rotation willow and poplar revealed an increase of soil C_{org} under the trees of 1300 kg ha^{-1} y^{-1}. By comparison, fertilization only accounts for a difference in soil C_{org} between fertilized and non-fertilized tree blocks of 250 kg ha^{-1} y^{-1}.

INTRODUCTION

Climate change and concern regarding future supplies of fossil fuels have led to growing interest in the potential of using biomass, particularly crops, for energy purposes. One environmental benefit of replacing fossil sources with biomass-based energy is that the energy obtained from crops does not add to the carbon dioxide (CO_2) increase in the atmosphere. However, energy from biomass is not CO_2-neutral because direct and indirect energy is required to grow and process crops. Additionally, due to fertilizer-induced emission of nitrous oxide (dinitrogen oxide: N_2O), the CO_2 advantage of biofuels diminishes and may vanish if exorbitant nitrogen fertilization is used for biomass production. A recent study (Crutzen et al 2007) reports that emissions from the burning of biofuels derived from rapeseed and corn have been found to produce more greenhouse gas emissions than they save. In practice, the advantages of reduced CO_2 emissions are more than offset by increased N_2O emissions. Furthermore, there is some discussion that short rotation coppice could lead to soil carbon losses.

The experimental field in Potsdam-Bornim/Germany, established in 1994, serves to provide precise experimental results regarding biomass yield of different crops on sandy soils and in the moderate climatic conditions of Central Europe. The influence of N_2O emissions on the CO_2 advantage of biofuels was to be followed up by means of systematic long-term measurement of N_2O emissions from the soil. And finally, the study was to check emissions from burning different kinds of biomass and the change in soil carbon content due to permanent and annual cropping.

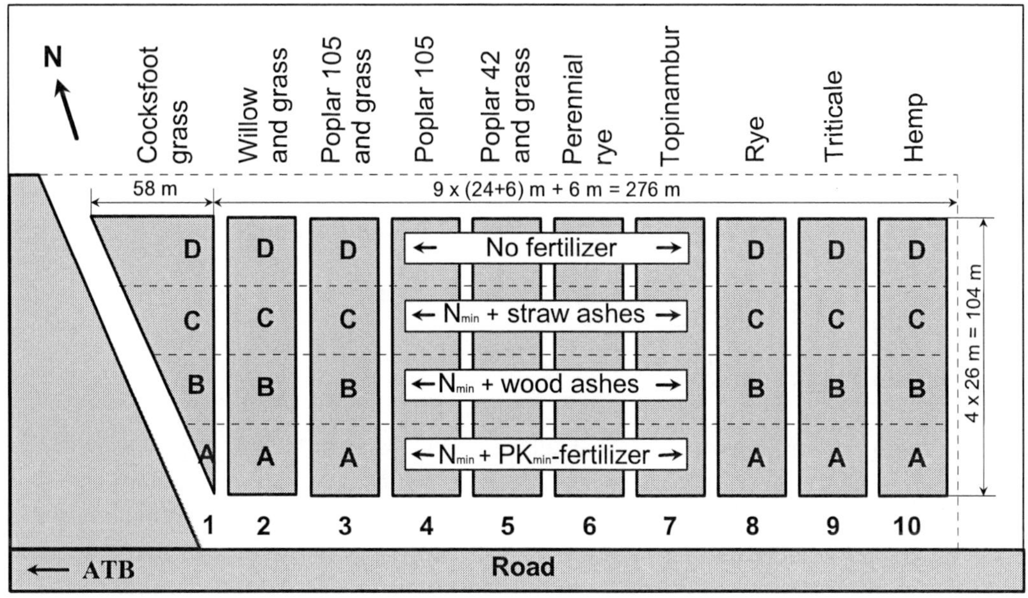

Figure 1. Experimental field of ATB (state 1996).

MATERIAL AND METHODS

Experimental Site

The experimental field was created in 1994. It is situated in the grounds of the Leibniz-Institute for Agricultural Engineering (ATB) northwest of Potsdam in the state of Brandenburg and has been studied right up to the present date. With a total area of 4.1 hectares (ha), it is part of a field which was previously cultivated in a conventional manner. The experimental field is divided into 10 long plots (1 - 10) of 0.25 ha each, which are subdivided into 4 blocks (A - D) of 624 m^2 each. Selected plots are further divided into a right (R) and a left side (L). Block A receives conventional basic mineral fertilization and 150 kg N ha^{-1}. On blocks B and C, wood and straw ashes as well as 75 kg N ha^{-1} each are applied. Block D is not fertilized. No pesticides have been applied anywhere on the entire area so far (Figure 1).

Fertilizers used comprise 540 or 270 kg ha^{-1} of calcium-ammonium nitrate and 524 kg ha^{-1} of a potash-phosphate mixture (100 kg K ha^{-1} and 26 kg P ha^{-1}), as well as grate ashes (coarse ashes) from wood and straw combustion plants. While the nitrogen fertilizer is applied on blocks B and C in 2 and on block A in 3 partial doses per year (in April and May), the basic fertilizers are applied only every other year. The mass of wood and straw ash on blocks B and C is adjusted to the amount of potassium (K) of the conventional basic fertilizer on block A, but did not exceed 3.3 t ha^{-1}. Because of the low potassium content of the wood ashes, it is necessary to apply additional K fertilizer on block B.

The soil of the experimental field is representative for the state of Brandenburg. With a land index of about 30, it is a relatively poor soil. In the upper horizons, down to a depth of 60 cm, weakly humus, slightly loamy sand and beneath this sandy loam of diluvian origin are predominant. The ground water level is about 8 m. At the beginning of the trial, the contents of organically bound carbon averaged 9.2 g kg^{-1} and the mean pH-value was 5.7 (Scholz et al. 1999).

The climate on the experimental site is temperate, however, with lower precipitation than in other regions of Germany. Between 1994 and 2007 the mean annual temperature was 9.6°C and the mean sum of precipitation 583 mm y^{-1}. Nonetheless, significant differences in yearly average temperature (7.3 to 10.5 °C) and precipitation (406 to 798 mm y^{-1}) occurred during this period.

Crop Species Investigated

Only crops that are appropriate for combustion or gasification are cultivated, with the focus on perennial plants such as cocksfoot grass, willow, poplar, miscanthus sinensis, perennial rye, and topinambur. Some of these species are conventional agricultural crops and known as modestly and environmentally kindly in cultivation and energetic use. Only few results and experiences are available in Europe for other species, especially short rotation coppice (SRC) and miscanthus sinensis. The varieties were selected upon the recommendation of various agricultural and forestry institutions (Table 1).

The short rotation coppices, poplar and willow, have been used and investigated over the full period of 14 years, while cultivation of the other perennial crops was stopped when the yield of block A fell below about 50 % of the first year's yield. The annual crops such as rye, triticale and hemp rotate on plots 8 to 10.

Table 1. Energy crops investigated

Plot		Species	Variety	Years
1	Cocksfoot	*Dactylis glomerata L.*	Lidacta	10
2	Willow	*Salix viminalis*	Salix 21	14
3	Poplar	*Populus maximowiczii x P. nigra*	Japan 105[1]	14
4	Poplar	*Populus maximowiczii x P. nigra*	Japan 105[1]	14
5	Poplar	*Populus maximowiczii x P. trichocarpa*	NE 42[2]	10
6	Miscanthus sinensis	*Miscanthus x giganteus*	--	2
6 - 7	Perennial rye	*Secale montanum L.*	Permontra	5
7	Topinambur[3]	*Helianthus tuberosus L.*	Parlow	7
6 -10	Winter rye	*Secale cereale L.*	Amilo	14
6 -10	Winter triticale	*Triticosecale Wittmack*	Alamo	9

[1] Similar to the multiple clone variety Max.

[2] Synonym for hybrid 275.

[3] Jerusalem artichoke, sunchoke.

Conventional farm machines are used for plowing, seeding and harvesting of the crops. After planting poplar and willow, a modest grass mixture (*Festuca rubra L.* and *Festuca ovina L.*, 1:1) was seeded an on plots 2, 3 and 5 in order to improve the biodiversity.

Measurement of Yield

The yield of the crops investigated is measured annually or biennially (SRC) at the traditional harvest time. Five areas of 10 m² each, located at least 2 m from the edge, are selected per block at random and harvested manually using a manual mowing-machine or a chain saw. That means that the whole crops are harvested, and haulm-type crops including weeds. The material is weighed with a balance (measuring error ± 100 g), and 5 samples of about 100 g dry matter (DM) each are taken to measure the moisture content in a drying oven at 105 °C.

Measurement of N_2O

Gas flux measurements were performed usually four times a week (Mon, Tue, Thur, Fri; between 10 a.m. and 12 noon) using closed chamber technique and a gas chromatograph (GC: Shimadzu GC 14A). The maximum capacity for the automated GC measuring system was 64 gas samples per measuring day. Eight gas samples are obtained from a measurement plot with four blocks A to D. Thus the decision was to include 6 plots (48 gas samples per measuring day) in the long term monitoring program and to use the remaining capacity reserve for control measurements on bare soil (four measuring rings) and for short-time tasks. PVC-sealing rings (Y profile, sealing by water level) for the gas flux chambers (cover boxes) were embedded in the soil of the measuring spots at blocks A to D of plot 2 (willow), of plot 4 (poplar Japan 105) and of two or three plots from 6 to 10 according to actual measuring needs. Willow (2) and poplar blocks (4) of the experimental field were arranged on two sides each (called "left" and "right"). The left side is harvested every four years. The right side has a shorter harvesting (rotation) period of two years. In order to monitor differences in N_2O fluxes caused by harvest and short rotation period, respectively, one of the poplar plots (plot 4 without undersown grass) has been established with two measuring rings at each block A to D of the "left" and "right" sides.

The PVC-gas flux chambers had a volume (V) to area (A) ratio of V/A = 0.315 m and a volume of 0.064 m³. Two evacuated gas samplers (100 cm³ bottles with Teflon sealing and vacuum taps) were connected to each box (Figure 2). The first was filled when the box was put on the water-sealed ring on the soil, and the second after one hour of enclosure time. The samplers were then connected with the PC-controlled GC injection system (Loftfield et al. 1997). For each level of fertilization, the emission factor of N_2O was calculated by taking the difference between the annual mean values of the fertilized blocks and the non-fertilized blocks. The emission factor, also called nitrogen conversion factor, is defined as the ratio of annual N_2O-N emission to annual input of fertilizer-N (De Klein et al. 2006).

Figure 2. Gas flux chamber on plant plot with measuring bottles ("gas mouse") and measuring ring on bare soil.

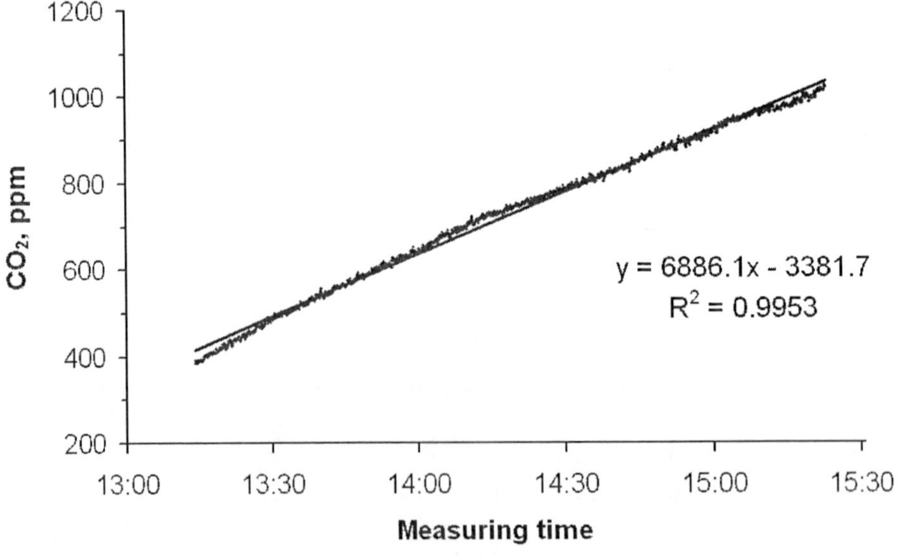

Figure 3. Concentration increase of CO_2 in a flux chamber on bare soil measured on May 16, 2007 with infrared CO_2 sensor (Model 950; Testo AG).

CO_2 and N_2O, both generated in the soil, have nearly equal diffusion constants. The easily measurable CO_2 served for the evaluation of linearity and mixing homogeneity of the measuring chamber (measurements at several heights in the closed chamber). Measurements of the concentration increase of CO_2 in flux chambers on bare soil showed that the increase in concentration was linear (coefficient of determination (R^2) higher than 0.99) during measurement periods (Figure 3).

Measurement of Soil Organic Carbon

To analyze the influence of the cultivation of energy crops on soil organic carbon (C_{org}) plot 2 (willow with undersown grass), plot 4 (poplar without undersown grass) and plot 8 (annual haulm-type crops) were sampled between August and October in the year 2006, i.e. 12 years after the establishment of the plantation. The crop rotation on plot 8 was mainly composed of rye and triticale, but included also hemp and will henceforth be referred to as 'annual crops'.

On the selected plots the blocks with the highest levels of fertilization (A) and with no fertilization (D) were sampled to estimate the influence of fertilization on the soil carbon stocks.

The soil was sampled to a depth of 30 cm with a root auger with an inner diameter of 8 cm, which allowed calculation of the soil density. On each tree plot 104 soil samples were taken (52 on block A and 52 on block D), while the sample number was halved on plot 8 (annual crops) where the spatial variability of soil C_{org} is lower due to regular tillage. Before the analysis, soil samples were dried and sieved through a 2 mm sieve. The remaining soil was weighed to calculate the soil density and afterwards sub-samples of 12 to 18 mg were prepared for the C/N-analysis by grinding and drying at 105° C.

Soil carbon concentrations were also part of the regular soil analysis conducted on the experimental field since the establishment of the plantation. For these analyses, five soil samples were taken on each block of each plot and mixed to form a composite sample. In 1994 these analyses were conducted twice, in April before the establishment of the plantation and in October shortly after the planting. Thus there exist 80 values of C_{org} concentrations for the first year. As soil density was not measured, however, accurate calculation of the initial carbon stocks is not possible. Furthermore, a statistical comparison with the carbon concentrations measured in 2006 is restricted, as the mixing of the samples reduced the variability. Despite these limitations, the soil C_{org} concentrations of 1994 were compared with those under annual crops in 2006 to test whether the carbon concentrations had changed in that time.

To estimate the influence of willow and poplar on soil carbon stocks, the measured C_{org} values of these two plots were compared statistically with the carbon stocks under annual crops. For the comparison, the non-parametric Mann-Whitney-U-Test and the Kruskal-Wallis-Test were used because the datasets are neither normal nor logarithmic normal distributed.

To test the influence of fertilization, the soil C_{org} stocks measured on the blocks A (conventional fertilization) were compared with those of block D (no fertilization) for each of the selected plots. For the comparison of the carbon stocks of blocks A and D of the willow plot the t-Test with Welch correction was used, as the datasets were logarithmic normal distributed but the variances were not homogenous (e.g. Lehn 2006, Fahrmeier 2001). Soil carbon stocks on the non-fertilized poplar block 4D were not normal distributed, so the datasets of blocks 4A and 4D were compared with the Mann-Whitney-U-Test. (e.g. Lehn 2006). On plot 8 (annual crops), the soil carbon stocks on blocks A and D were normal distributed and variances were homogenous, so the t-Test could be used for the statistical analysis. (e.g. Lehn 2006, Fahrmeier 2001). The statistical comparisons described in this study all tested the null hypothesis that the means of the compared statistical populations are equal. The statistical significance level α is defined as 0.05 for all tests. Accordingly the null hypothesis of a test is rejected when the p-value is smaller than 0.05.

Calculation of Energy Parameters

The calculation method applied to determine the energy input is based on the VDI 4600 guideline "Cumulated Energy Demand - terms, definitions, methods of calculation" (VDI 4600 1998). According to this, the expenditures of energy are divided in each process stage into energy input directly related to the process, and the indirect inputs, which can only be assigned by codes, for providing the implements, machinery and plant necessary for the process as well as for the conditioning of the surroundings. This means that the cumulated energy demand (CED) indicates the entire energetic input which results in relation to the production (P), the use (U) and the disposal (D) of all production facilities or services used:

$$CED = CED_P + CED_U + CED_D \tag{1}$$

As the relevant literature frequently differentiates between operation (O) (utilization) and maintenance (M) (repair, accommodation etc.) when presenting energy data for agriculture, the following specification is used:

$$CED_U = CED_O + CED_M \tag{2}$$

During the total "life cycle" of an operating supply used in agricultural production, certain quantities of various final energy sources such as coal, fuel and electricity are used. The ratio r of final energy to primary energy in each case is taken into account to calculate the cumulated energy demand (assessed in terms of primary energy). If the energy sources used cannot be distinguished, for reasons of simplification the mean ratio of $r_{mix} = 0.66$ for Germany stated in the national energy statistics is used for the calculation (Anonymous 2008).

In order to quantify the energy flows exactly, the balance scope must be clearly defined and delimited in terms of place, time and technology (Scholz & Kaulfuß 1995). In this case, the entire cumulated energy input of all the supply items used directly or indirectly in the production process under review are recorded. Human labor and other forms of energy present in the natural surroundings but not used technically are not included.

A special computer program based on the MS Access relational database system is used to calculate the cumulative energy demand (Scholz 1997). The $CED_{P,\,M\,or\,D}$ [MJ ha^{-1} y^{-1}] of production, maintenance or disposal of a machine, tool or any other equipment is calculated by

$$CED_{P,M\,or\,D} = ced_{P,M\,or\,D} * m * n * n_{Pr} * \Delta t_{Pr} / (\Delta T_{Ls} * \Delta T_{Pc}) \qquad (3)$$

where ced [MJ kg^{-1}] is the corresponding specific cumulated energy demand (Table 2), m [kg] the mass, n [-] the number, Δt_{Pr} [h ha^{-1}] the specific operating time and ΔT_{Ls} [h] the life span of the equipment used. n_{Pr} [-] is the total number of processes and ΔT_{Pc} [y] the total time of a production cycle of the calculated crop.

The CED_O of operation of a machine, tool or other equipment is calculated by

$$CED_O = \left(\sum_{i=1}^{m} ((H_{ui} + ced_{Pi})\, a_{Oi} * c_{Oi}) + \sum_{j=1}^{n} (ced_{Pj} * a_{Oj} * c_{Oj}) \right) n * n_{Pr} * \Delta t_{Pr} / \Delta T_{Pc}$$

$$(4)$$

with ced$_{Pi\,or\,Pj}$ [MJ kg^{-1}] as specific cumulated energy demand for production, $a_{Oi\,or\,Oj}$ [kg h^{-1}] as specific nominal consumption and $c_{Oi\,or\,Oj}$ [-] as consumption coefficient of the fuel i or the operational supplement j (Scholz, Berg & Kaulfuß 1998). H_u [MJ kg^{-1}] is the net calorific value of a fuel (Diesel: 42.7 MJ kg^{-1}; Gasoline: 43.5 MJ kg^{-1}). The nominal fuel consumption a_O of farm machinery can be found in literature, e.g. in the published database of KTBL (Anonymous 2004). The engine oil consumption is assumed to be 2.0% of this figure.

Table 2. Specific cumulated energy demand of production, maintenance and disposal of selected agricultural production utilities and fuels (mean values for Germany by Scholz, Berg & Kaulfuß (1998))

Production utilities and fuels		Specific cumulated energy demand		
		Production ced$_P$	Maintenance ced$_M$	Disposal ced$_D$
Four-wheeled tractors	MJ kg^{-1}	65	27	0,5
Self-propelled harvesters	MJ kg^{-1}	70	22	0,5
Trailed harvesters	MJ kg^{-1}	55	22	0,5
Application equipment	MJ kg^{-1}	55	15	0,5
Trailers	MJ kg^{-1}	50	25	0,5
Tillage equipment	MJ kg^{-1}	48	24	0,5
Nitrogen fertilizer	MJ kg^{-1}	50	-	-
Phosphorous fertilizer	MJ kg^{-1}	17	-	-
Potash fertilizer	MJ kg^{-1}	10	-	-
Lime	MJ kg^{-1}	3	-	-
Seeds (grain)	MJ kg^{-1}	6	-	-
Twine	MJ kg^{-1}	90	-	-
Engine oil	MJ kg^{-1}	54	-	-
Diesel	MJ kg^{-1}	7.2	-	-
Gasoline	MJ kg^{-1}	8.6	-	-

The $CED_{P \text{ or } D}$ of production or disposal of consumable materials non-related to machinery such as fertilizers, seeds, etc. is calculated by

$$CED_{P \text{ or } D} = \sum_{k=1}^{o} (ced_{Pk \text{ or } Dk} * a_{Mk}) * n_{Pr} / \Delta T_{Pc} \tag{5}$$

where a_{Mk} [kg ha^{-1}] is the specific consumption per process step of the material k.

The total cumulated energy demand is the sum of these CEDs of each process step. In order to guarantee the general applicability of the results, average practical conditions are assumed (field distance 2 km; field area 5 ha) corresponding to the soil quality of the experimental field described. All processes are included in the calculation, starting from preparation of soil up to harvesting.

The output of energy, hereinafter referred to as energy yield (EY), is calculated on basis of the fresh matter yield of biomass (Y_{FM}) and the net heating value (NHV) of each crop species.

$$EY = Y_{FM} * NHV \tag{6}$$

$$NHV = NHV_{DM}(1 - \frac{w}{100}) - 2.44\frac{w}{100} \tag{7}$$

where NHV_{DM} [MJ kg^{-1}] is the net heating value of the absolute dry material and w [%] the physical water content.

The energy gain (EG) is the difference between the energy yield and the cumulated energy demand.

$$EG = EY - CED \tag{8}$$

RESULTS

Yield

The yield is the most important result. It determines not only the economic but also the ecological and energetic efficiency. As known from conventional crops, the yield depends significantly on the fertilization level, particularly on the application of nitrogen.

On the intensively (i.e. conventionally) fertilized blocks A, the amount of nitrogen fertilizer is oriented to the average application rate for conventional crops in this region. Thus for reasons of comparative balances, this uniform rate is not always oriented exactly to crop demand. Nevertheless, the results allow general conclusions to be drawn.

The highest long-term yields on block A were achieved with hemp and poplar Japan 105 without undersown grass at 9.8 t_{DM} ha^{-1} and 9.9 t_{DM} ha^{-1} respectively, followed by annual and perennial rye, triticale, and cocksfoot at 8 to 9 t_{DM} ha^{-1}. The cereal yields correspond approximately to the average yields of these species in the region (Scholz et al. 2004;

Anonymous 2007). There are hardly any experiences and results in Europe regarding the perennial rye. Nevertheless, it is a promising energy crop species because it needs only one sowing for several harvests. Although the annual yield of the intensively fertilized blocks drops from 10.2 to 5.9 t_{DM} ha^{-1} within 3 years, the average yield of perennial rye is acceptable (Table 3).

The haulm of the originally promising topinambur plant, which is also called Jerusalem artichoke or sunchoke and which is a perennial species too, shows the lowest yield of all crops (4.2 t_{DM} ha^{-1} in 7 years). As the number of tubers and stems increases over the years, the annual yield decreases and the stability of the stems diminishes, which makes mechanical haulm harvesting more difficult. With 4,9 t_{DM} ha^{-1} y^{-1}, the mean 3-year yield of the high-fertilized block A is higher than the 7-year yield. Therefore it is advisable to terminate utilization of topinambur after a few years. However, in practice it is difficult to break off the cultivation without pesticides, because this native species of South America is very resistant. By the way, if the tubers, which e.g. can be used for ethanol production, are harvested too, the average haulm yield is only slightly less than indicated here. Just small harvest residues of tubers remaining in the soil are sufficient to re-establish this culture in the following year.

The average yields of the short rotation coppices (SRC) or 'field trees' are nearly in the range of conventional crops such as winter rye, which is the typical cereal in this region. Except for the poplar variety NE 42, which is not representative because it has an extremely high mortality rate, the mean dry matter yield of poplar and willow on the high-fertilized blocks A ranges between 8.1 and 9.9 t ha^{-1} per annum, depending on the variety and undersown crop. However, the annual yield of trees is very low during the first years but grows continuously up to a final value, which is reached after 2 to 8 years (Scholz & Ellerbrock 2002).

Table 3. Average long-term yield of the energy crops investigated (1994 - 2007)

Crop species	Years	Mean annual dry matter yield in t ha^{-1} y^{-1}			
		Block A 150 kg N ha^{-1} + min. fertilizer	Block B 75 kg N ha^{-1} + wood ash	Block C 75 kg N ha^{-1} + straw ash	Block D 0 kg N ha^{-1} no basic fertil.
Cocksfoot grass	10	8.0	7.2	7.3	5.4
Willow*,Salix 21	14	8.1	7.1	8.0	7.1
Poplar*, Japan 105	14	7.8	7.9	8.7	7.6
Poplar, Japan 105	14	9.9	10.0	9.5	10.1
Poplar*, NE 42	10	5,4	6.5	6.9	6.9
Perennial rye	5	8.5	8.0	7.4	6.1
Topinambur	7	4.2	4.1	3.9	3.3
Hemp	9	9.8	9.1	9.0	7.5
Winter rye	14	8.9	8.3	8.1	7.0
Winter triticale	9	8.3	8.0	7.8	6.0

* With undersown grass.

The undersown crop, a modest grass mixture, proves to be a significant water and nutrient competitor, which causes the yield of the balsam poplar hybrid Japan 105 to diminish by 10 to 65% within the first 4 years depending on the fertilizing regime. In contrast to willows, however, the large leaves of poplars suppress the undersown crop (and weeds) in the course of time so that yield differences diminish considerably over time. Therefore it seems to be inadvisable to sow other crops on poplar fields the more so as the census of arthropods – an indicator of faunistic variety - shows no difference between plots with and without undersown grass (Scholz et al. 1999).

Even though pesticides were consistently not used, pest infestation and plant diseases remained within limits and did not cause any detectable yield depressions. Since weeds are usually harvested with the whole (haulm-type) crops, yield losses by comparison with a weed-free culture are insignificant, as was proven on other sites (Karpenstein-Machan 2000).

With respect to environmental and energetic aspects, it is necessary to know to what extent the fertilizer can be reduced without reducing the energy efficiency. Results obtained with conventional cereals are only transferable to energy crops to a limited extent because the investigations of food crops were mostly focused on grain, but not on whole plants, and there are so far only very few and very limited results available for cultivation of poplars and willows on farm land.

As regards the impact of nitrogen fertilization on the development of yields, the differences between haulm-type and woody crops are evident. The yields of all less and non-fertilized haulm-type crops such as grass, rye, triticale, hemp and topimanbur decrease over the years.

Although the statistical reliability - due to the yearly changing weather conditions - is very limited, the data permit some general predictions. In relation to a nitrogen application of 150 kg N ha^{-1} (block A), the yield of the investigated haulm-type crop species decreases by only 0.5 to 2.0% p.a. at 75 kg N ha^{-1} (average of block B and C during $\leq$ 14 years), where the two cereals show the lowest losses. This means that these crop species can guarantee relatively high yields at the present location, even if nitrogen supply is reduced for many years. Complete omission of fertilization (block D) leads to a long-term yield reduction of approximately 2 to 7% per annum. The extreme decrease in the yields of cocksfoot grass and topinambur haulm is caused not only by lack of nitrogen but - as was proven by soil analyses - also by lack of other nutrients, particularly potassium (Figure 4, top).

The yields of poplars and willows are determined less by the fertilizers than by the undersown crop and the age of the trees, or more precisely of the roots. A reduction of fertilizer causes a loss of yield only during the first few years. So e.g. in the first four years the yield of non-fertilized willow (block D) is only 50% related to the highly fertilized block (A). However, in contrast to haulm-type crops the relative yield of the less and non-fertilized trees does not diminish over the years but increases by approximately 0.5 to 4% per annum. After 10 years there is nearly no difference between the three fertilization levels and the yield amounts to roughly 100% or more (Figure 4, bottom).

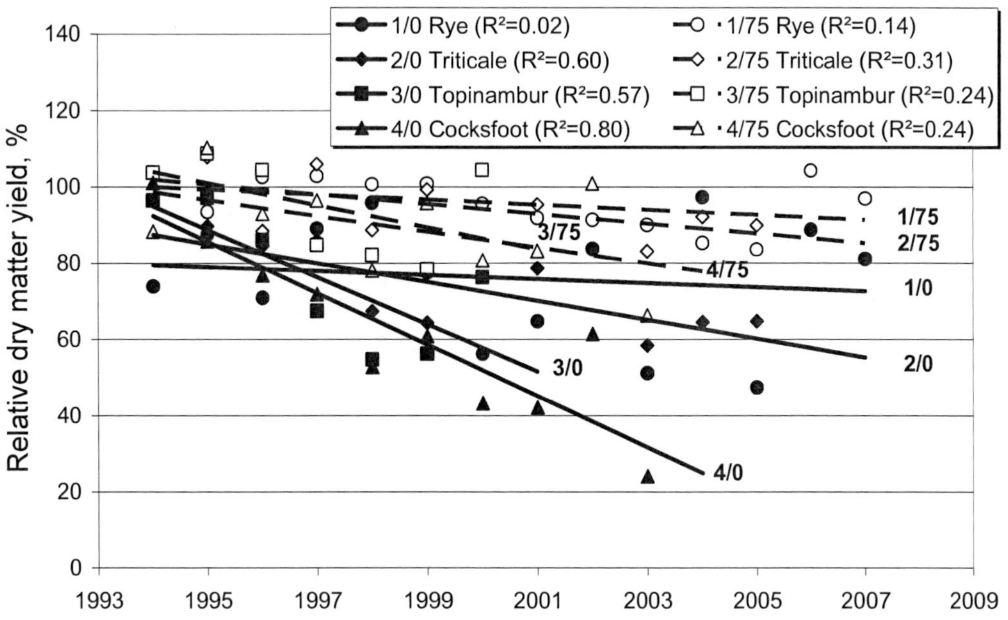

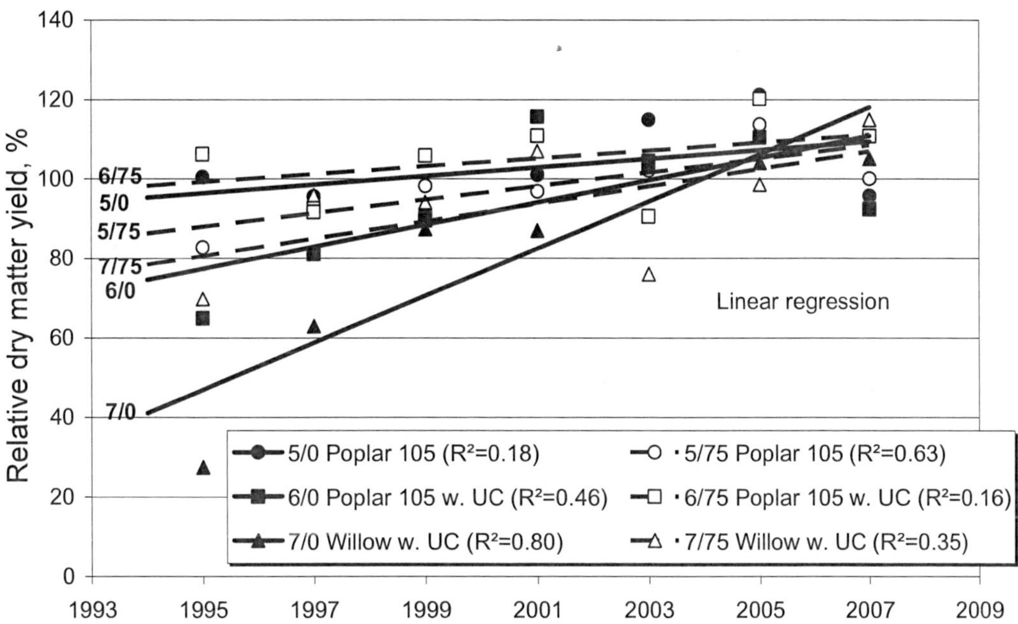

- - - - - Regression line of 75 kg N ha⁻¹ and wood or straw ash (average of blocks B and C).
-------- Regression line of zero fertilization (block D).
UC... Undersown crop (grass).
R²..... Coefficient of determination.

Figure 4. Impact of reduced fertilization on the development of yields of haulm-type crops (top) and short rotation coppices (bottom) in relation to the yields of conventionally fertilized crops.

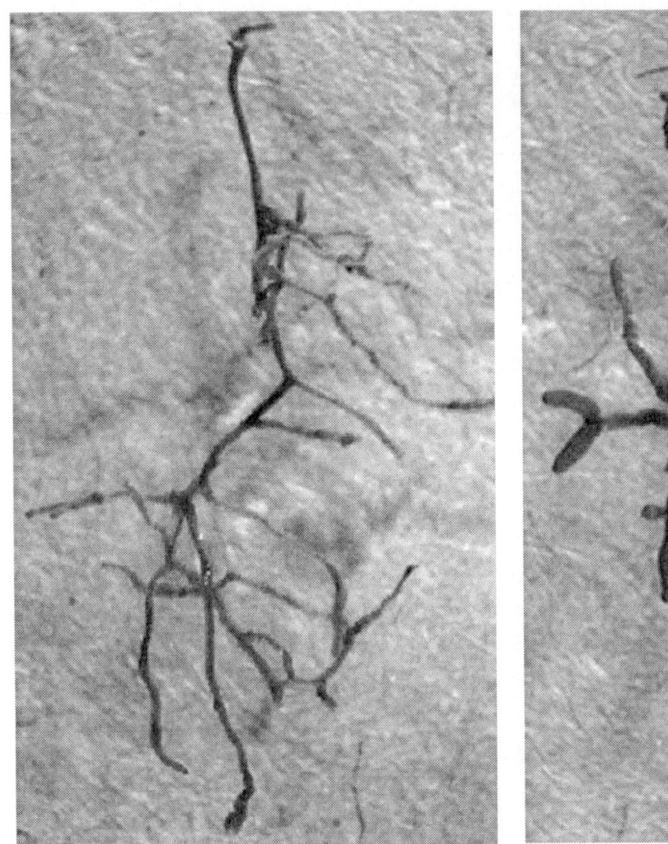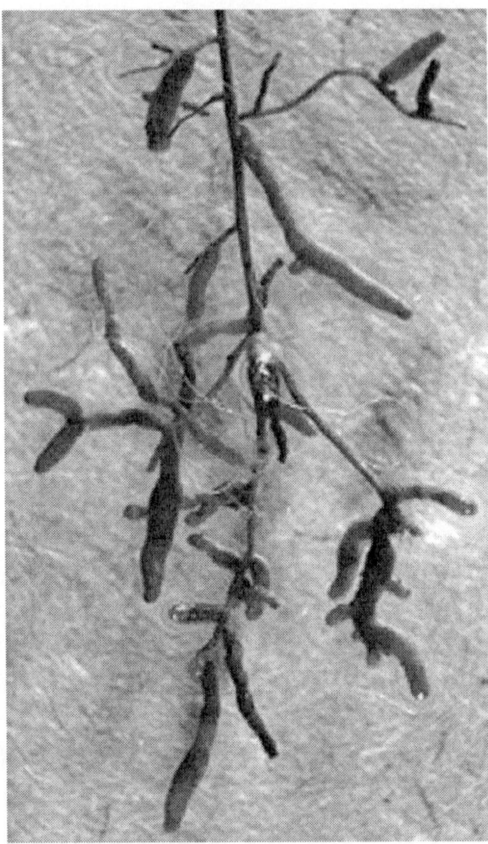

Figure 5. Mycorrhica on poplar roots of the highly fertilized plot A (left) and of the non-fertilized block D (right). (Photo: Forstreuter M., TU Berlin)

The reasons for the time-dependent approximation of the yields of differently fertilized SRC may be both a reduction of the yields on highly fertilized blocks due to the fertilizer-caused growth of weeds (particularly during the first years) and an increase of yields on less or non-fertilized blocks as a result of extending roots, which utilize the nutrients of lower soil layers, as well as of extending mycorrhiza. This symbiosis of fungi and plants on fine roots supplying crops with additional nutrients and water is supposed to be the main reason for the nitrogenous modesty of short rotation coppices (Jug 1998; Bungart 1999). A microscopic analysis of fine roots on blocks A and D seems to confirm this thesis. The measured rates of mycorrhiza on non-fertilized poplar roots are much higher than on highly fertilized roots, which is an indication that poplars are able to compensate the lack of nutrients by means of such a symbiosis (Figure 5).

Energy Gain

The energy gain - comparable with the economic gain - is the difference between cumulated energy demand (input) and energy yield (output). Thus the cumulated energy demand (CED) calculated using the method described above has a relevant impact on the energy gain and the energetic efficiency of the production of energy crops. On conventionally fertilized fields corresponding to block A of the experimental field, the cumulated energy demand for the cultivation and harvesting of crops examined here amounts to approximately 14 to 17 GJ ha^{-1} y^{-1}. As a general rule, the energy requirements per year diminish as production cycles and harvesting intervals become longer, so that perennial cultures generally require less energy. Post-harvest processes such as storage, drying and transport are not considered in these figures. Depending on culture and technology, they would need a further 2 to 6 GJ ha^{-1} (Scholz & Kaulfuß 1995).

In view of the energy-intensive production of nitrogen fertilizer (ced$_P$ = 50 MJ kg^{-1} N), reducing its use significantly affects the total energy demand. Not only is the CED of fertilizer production dispensed with, but also the CED of its application. Compared with an application rate of 150 kg N ha^{-1} y^{-1}, and conventional basic fertilization, in the case of no fertilization the total CED diminishes so that without fertilization only 2.2 to 4.9 GJ per ha and per year are needed for cultivation and harvesting of the crop species examined. The extremely low value for short rotation coppice results from the assumed long production cycle of 20 years, as well as from the long harvesting interval of 4 years.

The energy yield (EY) calculated on the basis of yield, calorific value, and water content of the harvested crops, is particularly dependent upon the species, the undersown crop, and the fertilization. If those species whose yield is extremely low, such as topinambur haulm, intensively fertilized poplar NE 42 and non-fertilized cocksfoot grass, are disregarded, the energy yield ranges between 100 and 160 GJ ha^{-1} y^{-1} corresponding to an oil equivalent of approximately 2,600 to 4,300 liters per hectare and per year (Table 4).

Table 4. Energy demand (CED) and yield (EY) of energy crops depending on fertilization

Crop species	Net heating value (MJ kg$_{DM}$$^{-1}$)	Energy in GJ ha^{-1} y^{-1}					
		Block A 150 kg N ha^{-1} + min. fertilizer		Block B+C 75 kg N ha^{-1} + biomass ash		Block D 0 kg N ha^{-1} no basic fertil.	
		CED	EY	CED	EY	CED	EY
Cocksfoot grass*	16.5	15.4	128.5	8.3	116.5	3.0	86.7
Willow**[1], Salix 21	18.3	14.5	128.5	6.3	119.7	2.2	112.6
Poplar**[2], Japan 105	18.4	14.5	124.5	6.3	132.5	2.2	121.3
Poplar**, Japan 105	18.4	14.5	158.0	6.3	155.6	2.2	161.2
Poplar**[1], NE 42	18.4	14.5	86.2	6.3	106.9	2.2	110.1
Perennial rye*	17.2	15.2	142.4	8.1	129.0	2.9	102.2
Topinambur*	18.0	15,5	73.7	8.4	70.2	3.1	57.9
Hemp*	17.0	17.2	162.3	10.1	149.9	4.8	124.2
Winter rye*	17.2	17.3	149.1	10.2	137.4	4.9	117.3
Winter triticale*	17.1	17.3	138.3	10.2	131.6	4.9	100.0

* water content of harvested haulm-type crops w = 15%.
** water content of harvested woody crops w = 50%.
[1] cultivated with undersown grass.

The ratio of energy demand and yield (input/output) characterizes the energetic efficiency of the production of energy crops. With the exception of topinambur, poplar NE 42 and cocksfoot, this ratio amounts to 0.01...0.13. This means that only 1 to 13 % of the crop energy content must be expended for their cultivation and harvesting. The most favorable, i.e. the lowest value is achieved with the poplar variety Japan 105 without undersown crops and fertilization.

However, by contrast with other renewable energy sources, the decisive criterion in the case of energy crops is energy gain rather than the input/output ratio, because the availability of agricultural land is limited. Independently of the fertilization level, the annual (net) energy gain which results from the difference between energy demand (CED) and yield (EY) ranges between 95 and 145 GJ ha^{-1} y^{-1} for whole crop cereals and hemp. With 143 to 159 GJ ha^{-1} y^{-1}, the poplar variety Japan 105 without undersown crops also achieves rather high energy gains. The energy gain of the variants without fertilization is particularly high.(Figure 6).

The differences in energy gain between intensive and reduced fertilized haulm-type crops are no more than 10%. This means that acceptable energy gains can be achieved with approximately half of the N fertilizer which is usually applied on conventional food crops. A nitrogen application rate in excess of this quantity results in only a slightly higher energy gain, and for short rotation coppices even in a loss of gain, and is therefore energetically inefficient and in addition harmful to the environment.

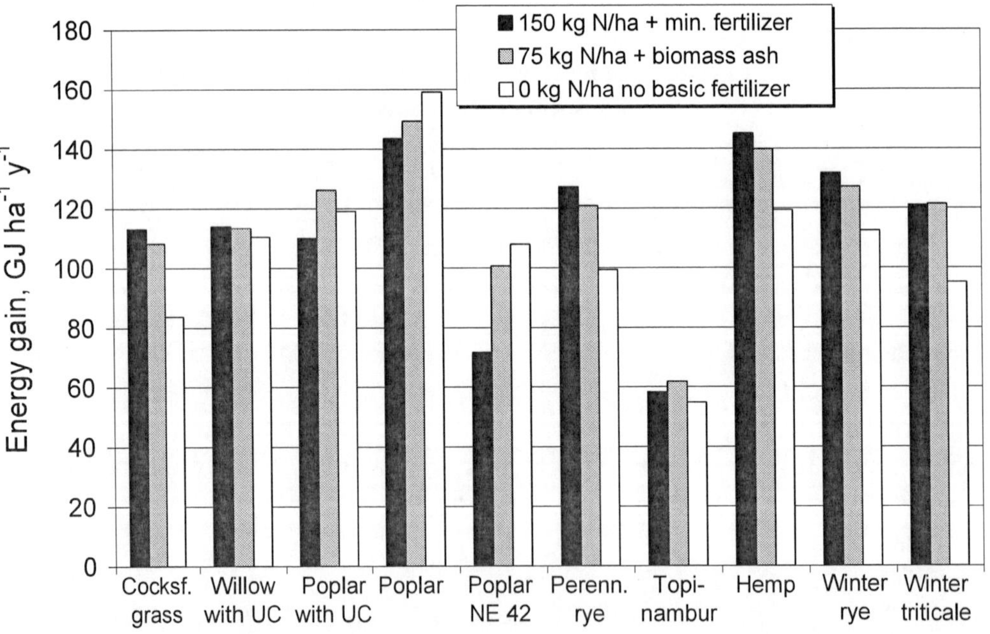

UC: Undersown crop (grass).

Figure 6. Energy gain in the production of the investigated energy crops under practical conditions depending on the fertilization level.

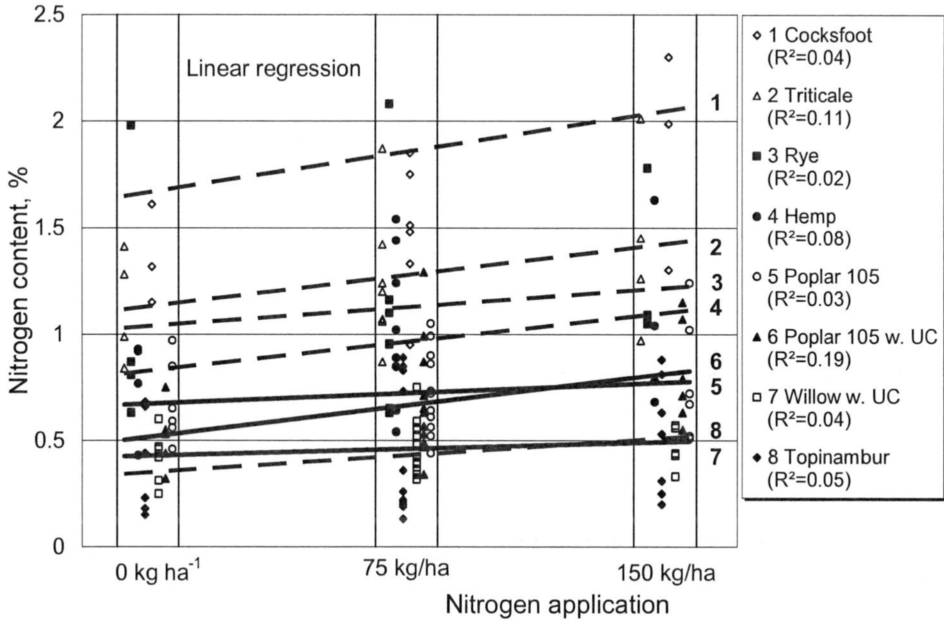

- - - haulm-type crops. ------ woody crops (SRC).

Figure 7. Nitrogen content of energy crops versus the nitrogen fertilization rate.

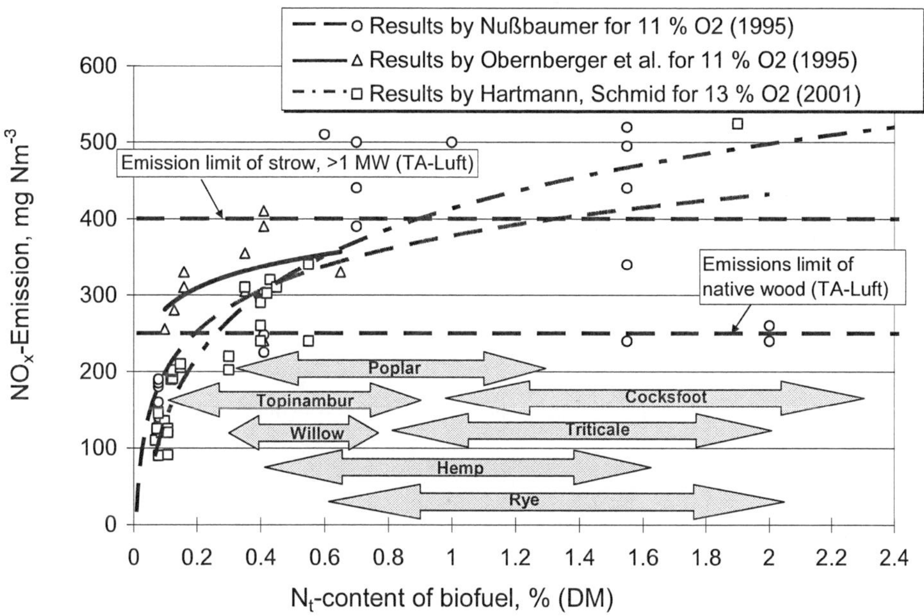

Figure 8. Nitrogen oxide (NOₓ) emissions during combustion versus nitrogen content of energy crops

Nitrogen Content and NO_x Emissions

The environmental relevance of nitrogen results not only from the effects of the fertilizer on plant, soil, water and air during the cultivation of energy crops, but also from the emissions during their combustion or gasification. Hence the content of nitrogen (N_t) in the crops is an important ecological parameter. In contrast to most food crops, low contents in energy crops are favorable.

The nitrogen content of the investigated crops shows a very wide span. The extreme values of a total of 160 analyses range between 0.1 and 3.3%. This span may be partly caused by sampling and/or analyzing errors, however. In literature the span of these crop species is a good deal wider (Scholz, 2004). With 2.1% the cocksfoot grass has the highest mean value (block A), while topinambur haulm has the lowest with 0.3% (block D). The nitrogen contents of poplar and willow are also very low, and the contents of hemp, rye and triticale lie between those of the trees and the grass.

The results of these analyses enable a co-relation to be established between the amount of fertilizer and the content of nitrogen, which is confirmed by means of a statistical regression analysis. So the application of 150 kg N ha^{-1} causes an average increase in the N_t content of approximately 0.1% to 0.4%, depending on the crop species (Figure 7).

As proven by Nussbaumer (1997), Obernberger (1997) and Hartmann & Schmid (2001), high nitrogen contents in biofuels induce high emissions of nitrogen oxide (NO_x) during combustion in up-to-date heating or gasification systems. Taking into account, that an absolute increase of 0.2% N_t results in an increase of 10 to 100 mg NO_x per 1 m^3 of exhaust fumes (under standard conditions), the application of 150 kg N ha^{-1} causes additional NO_x emissions in this range induced only by the N fertilization. Or in other words, the combustion of conventionally fertilized cocksfoot grass causes about 200 mg m^{-3} more nitrogen oxide than the combustion of non-fertilized poplars, which is not insignificant given legal emission limits of 250 to 400 mg m^{-3} (Figure 8).

Fertilizer-induced N_2O Emissions

Nitrogen fertilizing is one of the main sources of anthropogenic contribution to global N_2O emission. In soil, N_2O is produced predominantly by two microbial processes, the oxidation of ammonium (NH_4^+) to nitrate (NO_3^-) and the reduction of NO_3^- to gaseous forms NO, N_2O, and N_2 (Firestone 1982). The rate of N_2O production depends on the availability of mineral N in the soil and the conversion factor depends on soil type and climate (e.g. Bouwman 1990, 1996; Bouwman et al. 2002; Granli & Bøckman 1994; Stehfest & Bouwman 2006, Novoa & Tejeda 2006). For N_2O inventories, a default value of 1% (with additions and subtractions dependent on kind of fertilizer, soil type and climate) is recommended for mineral fertilizers (De Klein et al. 2006). When the cultivation of crops is assessed with regard to greenhouse gas abatement, this conversion factor plays a significant role, e.g. the nitrogen fertilizer-induced emission of N_2O-N may counterbalance the CO_2 advantage of biofuels (in cases of high nitrogen fertilizer application and conversion factor greater than 2% (Crutzen et al. 2007)), since N_2O as a greenhouse gas contributes to global warming 298 times more effectively than CO_2 (IPCC 2007).

Agronomic practices such as tillage and fertilizer applications can significantly affect the production and consumption of N_2O because of alterations in soil physical, chemical, and biochemical activities. An increase in N_2O flux rates following N-fertilizer applications has been observed in field and laboratory experiments (e.g. Jackson et al. 2003; Kaiser et al. 1998; Mulvaney et al. 1997). N_2O emission from croplands at site scales occurs essentially with great spatial and temporal variability (Dobbie & Smith 2003; Hellebrand et al. 2003, 2005; Veldkamp & Keller 1997). The annual pattern of temporal variation of N_2O emissions is determined in the temperate regions by the seasons and weather conditions, since soil N_2O emissions are regulated by temperature and soil moisture and so are likely to respond to climate changes (Frolking et al. 1998; Ruser et al. 2006).

The emission of N_2O followed the expected pattern with low emissions during the winter season, highs during the vegetation time and after fertilization. Enhanced N_2O emissions were detectable at fertilized blocks after fertilizing and lasted from three to six weeks. We also found temporarily and spatially limited high fluctuations throughout the entire study since 1999. N_2O emission peaks over 1000 µg N_2O m^{-2} h^{-1} were observed from a few measuring spots (Figure 9).

These findings are in accordance with other studies (e.g. Augustin et al. 1998; Röver et al. 1998). Because measurements were taken four times a week, the emissions could be studied with sufficient temporal resolution at different crop sites. Except during the few freeze-thaw cycles in the course of the winter season, the N_2O emission rate usually dropped to less than 30 µg N_2O m^{-2} h^{-1} between October and March.

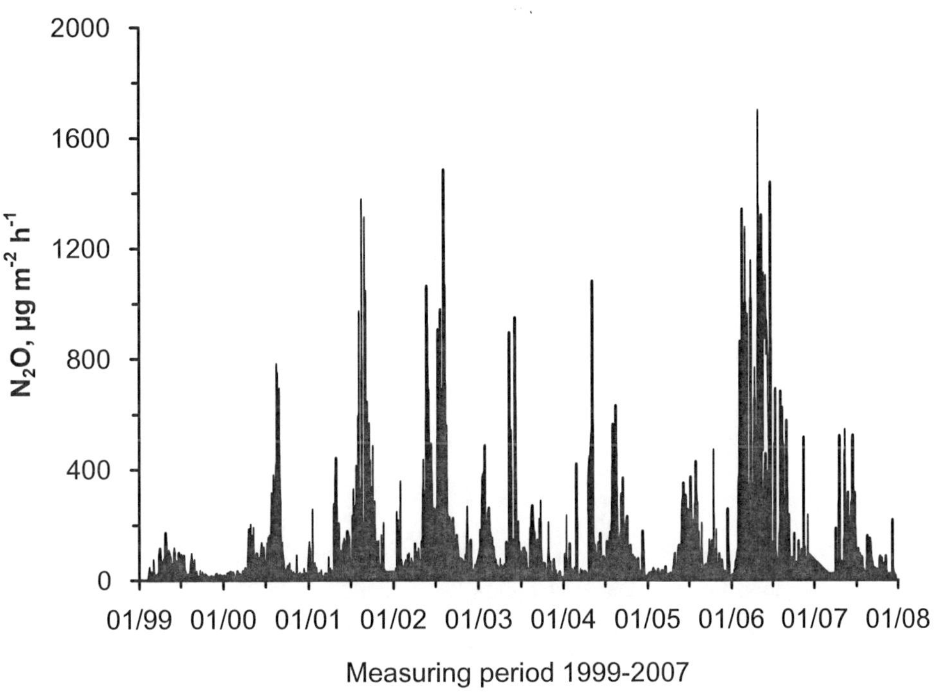

Figure 9. Time series of N_2O emissions from all blocks of the experimental field.

No clearly frost-induced emission rates were found during the winter seasons of 1999/2000, 2004/2005 and 2006/2007. Frost-induced N_2O-emissions are related to the course of temperature and could be dependent on several factors, such as groundwater level, water-filled pore space, soil pH, soil nitrate content, soil texture, and soil structure (Flessa et al. 1998; Mogge et al. 1996; Röver et al. 1998; Teepe et al. 2001, 2004). However, the frost-induced emissions we measured at a couple of the blocks are small compared with the total annual N_2O emission budget from the sandy soil of the experimental field. Therefore, these emissions were not excluded in the calculation of the nitrogen conversion factor.

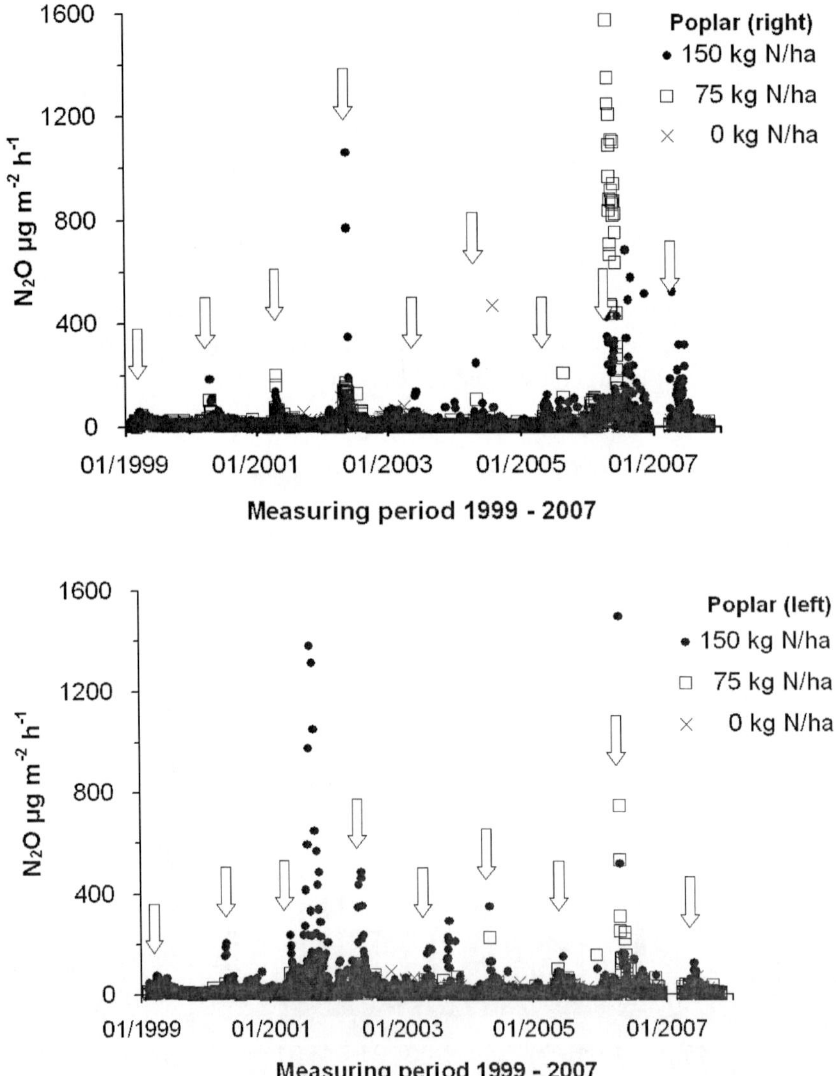

Figure 10. Time series of N_2O emissions from poplar plot 4 of the experimental field. Fertilizer-induced N_2O emissions are indicated by arrows (⇩).

There were several high emission periods at fertilized plots with emission rates of more than 1000 µg N_2O m^{-2} h^{-1} (Figure 9 and 10). Longer-lasting high N_2O emissions, called "hot spots" or "hotspots" (e.g. Christensen et al. 1990; Röver et al. 1999; Wanga et al. 2006), were detected at fertilized blocks only. N_2O hot spots observed here may generate up to 3 g N_2O m^{-2} y^{-1} (which corresponds to 20 kg N_2O-N ha^{-1} y^{-1} in other units) in the course of few months. Thus they can cause a local measured emission factor of more than 10 % at the corresponding measuring ring. The reason for these N_2O hot spot emissions is not clear. High emissions after harvesting were observed several times and might be connected with soil distortions.

The N_2O nitrogen conversion factor is defined as N_2O-N emission caused by fertilization in relation to the nitrogen fertilizer applied. The emission period considered is one year and the fertilizer-caused emissions are obtained by taking the difference between fertilized and non-fertilized blocks (Bouwman 1996). Different results will be obtained depending on whether N_2O hot spots are included or excluded. At first sight it appears reasonable to exclude N_2O hot spots from the calculation of the conversion factor, since the sources of these N_2O hot spots are unknown and do not seem to be directly be connected with fertilization level. On the other hand, N_2O hot spots were observed at fertilized blocks only. Thus these hot spots might have been caused indirectly by fertilization (in a presently unknown fashion) and, therefore, should be included into the determination of emission factors (Tables 5 and 6).

The so-called background flux (block D – without fertilization) is in the range of 0.5 to 1.5 kg N_2O-N ha^{-1} y^{-1}. This annual flux is typical for cultivated land with sandy-loamy soils and temperate climate (Bouwman 1996; Bouwman et al. 2002; Stehfest & Bouwman 2006). Except for rye (perennial rye and winter rye), the annual N_2O-N fluxes of fertilized blocks are below 2 kg N_2O-N ha^{-1} y^{-1}. This results in moderate emission factors (Table 6).

Table 5. Crop-specific N_2O-N emissions. Data evaluation with and without N_2O hot spots. Mean values of corresponding crop plots and years.

Crop species	Plot	Years	Mean annual N_2O-N emissions in kg N ha^{-1} y^{-1}					
			Plot A 150 kg N ha^{-1}		Mean of B, C 75 kg N ha^{-1}		Plot D 0 kg N ha^{-1}	
			without HS	with HS	without HS	with HS	without HS	with HS
Cocksfoot grass	1	3	1.30	1.30	1.10	1.10	1.06	1.06
Willow[1]	2	9	1.15	1.15	0.94	0.94	0.57	0.57
Poplar[2]	4L	9	1.57	2.26	0.81	0.89	0.53	0.53
Poplar[2)]	4R	9	1.47	1.72	0.87	1.39	0.47	0.47
Winter rye	6 - 10	8	2.43	2.82	1.55	1.55	0.94	0.94
Perennial rye	5 - 7	7[3]	4.02	4.19	2.23	2.23	1.30	1.30
Winter triticale	6 - 9	4	1.91	2.10	1.45	1.57	0.74	0.74
Hemp	6, 8	2	1.29	1.29	0.89	0.89	0.67	0.67

HS: N_2O hot spot.
[1] with undersown grass.
[2] harvest intervals: 4 years on 4L and 2 years on 4R.
[3] plot 5R in 2005-2007; plots 6 and 7 in 2006 and in 2007.

Table 6. Crop-specific N₂O-N nitrogen conversion factors and standard deviations for data evaluation with and without N₂O hot spots. Mean values of all A, B, and C blocks and years.

Crop species	Plot	No. of blocks	with N₂O hot spots		without N₂O hot spots	
			Mean value	Standard deviation	Mean value	Standard deviation
Cocksfoot grass	1	9	0.09	0.21	0.09	0.21
Willow[1]	2	27	0.45	0.48	0.45	0.48
Poplar[2]	4L	27	0.71	0.92	0.51	0.57
Poplar[2]	4R	27	1.07	2.21	0.60	0.46
Perennial rye[3]	5 - 7	21[3]	1.37	0.88	1.34	0.85
Winter rye	6 -10	24	0.97	0.56	0.88	0.45
Winter triticale	6 - 9	12	1.04	0.68	0.89	0.53
Hemp	6 - 8	6	0.33	0.18	0.33	0.18
Mean of all crops			0.83	1.15	0.68	0.63

[1] with undersown grass.

[2] harvest intervals: 4 years on 4L and 2 years on 4R.

[3] plot 5R in 2005-2007, plots 6 and 7 in 2006 and in 2007.

Table 7. Annual N₂O-N nitrogen conversion factors for two different fertilization levels with and without N₂O hot spots

Year	N₂O-N conversion factor in %			
	with N₂O hot spots		without N₂O hot spots	
	150 kg N ha⁻¹	75 kg N ha⁻¹	150 kg N ha⁻¹	75 kg N ha⁻¹
1999	0.26	0.19	0.26	0.19
2000	0.34	0.08	0.34	0.08
2001	1.19	0.36	0.68	0.36
2002	1.36	0.95	1.19	0.95
2003	0.52	0.60	0.47	0.51
2004	0.79	0.78	0.70	0.78
2005	0.80	0.62	0.61	0.59
2006	2.05	2.28	1.67	1.33
2007	1.12	0.70	1.12	0.70
99-07	0.94	0.73	0.78	0.61

Willow and poplar plots were measured throughout the whole period 1999-2007, whereas data from grass, rye, triticale, and hemp plots were not collected every year. The N₂O emissions from hemp plots were measured in 2000 and 2005 only. In these years the levels of N₂O emissions were low (2000) and moderate (2005) (Figure 9). No hot spots occurred in 2000 and only weak hot spots in 2005. Thus a hemp-specific emission value based on N₂O flux values resulting from these two years alone cannot be stated. The N₂O emissions depend on the course of the weather (e.g. in general we have lower N₂O emissions in dry years and increasing N_2O emissions after moist weather periods). Therefore it is difficult to compare plant-specific emissions, taking into consideration the high local and temporal variability of N_2O emissions. The comparatively high flux rates of perennial rye may be partially explained by this effect. The measurements of perennial rye plots 6 and 7 cover two years (2006 and 2007), and those of perennial rye plot 5R cover three years (2005-2007). The year 2006 produced high fluxes, presumably due to the unusually long snow period in Central Europe from mid-December till mid-April. In 2006, the N_2O emissions were higher compared with all other years (Figure 9 and Figure 10), followed by increased nitrogen conversion factors in 2006 (Table 7). Additionally, the soil of perennial rye plots (as of all other plots too) was neither cultivated nor sprayed, with the outcome of comparatively high weed density. It is not clear whether weeds could have contributed to increased N_2O emissions. Only legumes are known for nitrogen storage in soil and resulting production of enhanced N_2O emissions (Van der Weerden, 1999).

The comparison between poplar sides left and right indicates that there is no clearly measurable influence of the crop rotation period on N_2O emissions. Blocks D are nearly identical and the results without hot spots showed no significant difference between the fertilized blocks either. There was one hot spot of poplar left block A in 2001, and one other hot spot was observed on block B of poplar right in 2006. As the intensity of these two hot spots differed, the conversion factors calculated with hot spots differ from each other (Table 6), whereas the exclusion of hot spots in the calculation of the conversion factors results in comparable factors.

There is an obvious difference between the nitrogen conversion factors if N_2O hot spots are excluded, especially in the years 2001 and 2006 (Table 7). It should be noted that difficulties arise if the conversion factor of different crop species is to be evaluated using data from different years. One main problem for the comparison is the variability of the weather conditions, which causes different N_2O emission levels in dependence on precipitation and temperature of the year considered. In addition, there are hints that soil cultivation influences the N_2O emission level too. Changes in physical structure by soil tillage may alter biological activity and thus N_2O emissions over the crop season (Jackson et al. 2003; Kaiser et al. 1998; Kaiser & Ruser 2000; Mulvaney et al. 1997). All in all, the nitrogen conversion factors found here do not cause a real CO_2 equivalent burden for energy crops produced on sandy loamy soils.

Soil Organic Carbon

Global estimations denote that 500 Pg (1 Pg = 10^{15} g) organic carbon are stored in herbal biomass, and even 1500 Pg in the top meter of the soils (Scheffer & Schachtschabel 2002). With this huge amount of stored carbon (C), soils play an important role in the global carbon cycle, especially as soil carbon pools are not stable but influenced by vegetation, climate (precipitation and temperature) and other environmental factors, and hence may act as a source or a sink for CO_2. The vegetation has a great impact on soil organic matter, directly through the input of plant material (above and below ground), but also through the influence on temperature (shading), water content of the soil, and protection against erosion. Generally, soils under grassland and forests contain more carbon than adjacent arable land due to higher litter input and no-till of the soil (Scheffer & Schachtschabel 2002).

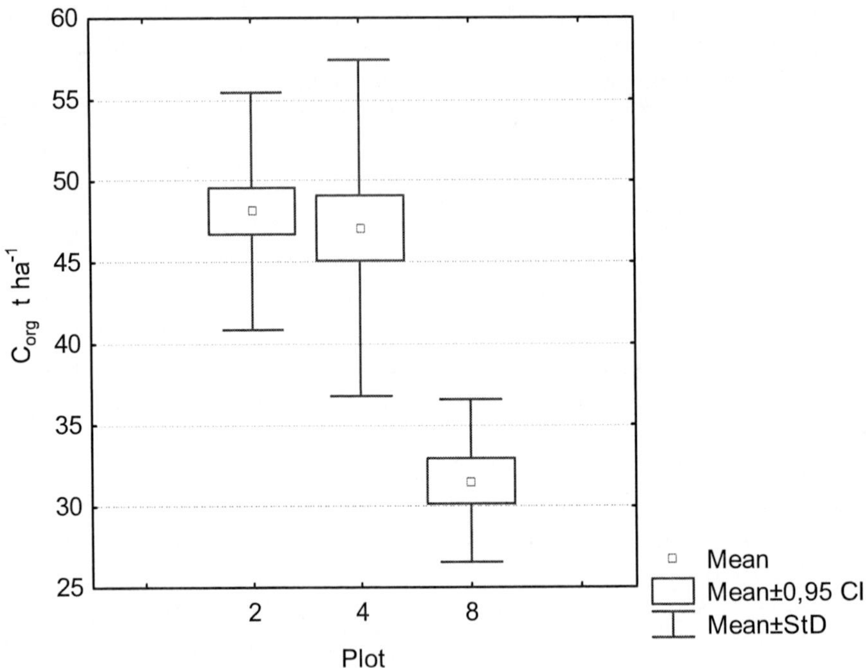

CI: Confidence interval. StD: Standard deviation.

Figure 11. Box-Whisker-Plots of the carbon stocks (depth 0-30 cm) for willow (plot 2), poplar (plot 4) and annual crops (plot 8) in 2006.

This could also be proved on the experimental field. The analysis of soil carbon stocks under willow (plot 2), poplar (plot 4) and annual crops (plot 8) with the Kruskal-Wallis ANOVA detected a significant difference between these three groups (Figure 11). The pairwise comparison of the medians of the carbon stocks with the Mann-Whitney-U-Test showed the difference between willow and poplar as not significant ($p = 0.183$), whereas soil C_{org} stocks differed significantly between each of the two tree species and annual crops ($p < 0.0001$ for both comparisons).

Assuming that the carbon stocks did not change in the soil under the annual crops in the 12 years, a mean annual carbon sequestration rate of 1300 kg ha^{-1} can be calculated for the two tree species. However, the statistical analysis revealed that the carbon concentrations under annual crops decreased between 1994 and 2006 from 9.2 g kg^{-1} (0-25 cm depth) to 7.0 g kg^{-1} (0-30 cm depth). Hence it must be noted that the calculated rate of 1300 kg ha^{-1} y^{-1} is composed of a decrease under annual crops and an increase under short rotation coppices. This could explain why this rate is relatively high by comparison with other studies (Grigal & Berguson 1998; Stetter & Makeschin 1999; Ulzen-Appiah et al. 2000). It must be noted, however, that Kahle and Boelcke (2004) detected comparable carbon sequestration rates on two sites in north-eastern Germany.

Regarding the impact of fertilization, higher soil carbon levels are observed on blocks A (conventional fertilization) than on blocks D (no fertilization). However, the statistical analysis detects that this effect is only significant for annual crops ($p < 0.0001$) and poplars ($p = 0.0093$), but not for willows ($p = 0.3433$).

Table 8. Means and standard deviations of soil carbon stocks under willow, poplar and annual crops with and without fertilization in 2006

		Organic carbon in soil under					
		Willow (2)		Poplar (4)		Annual crops (8)	
		Bl. A	Bl. D	Bl. A	Bl. D	Bl. A	Bl. D
Mean	t ha^{-1}	49.05	47.26	49.18	45.01	34.74	28.36
Standard deviation	t ha^{-1}	8.67	5.53	9.10	11.12	4.13	3.58
Sample number	-	52	52	52	52	26	26

Block A: 150 kg N ha-1 a-1 and mineral basic fertilizer over 12 years.
Block D: no fertilization over 12 years.

The mean increase of C_{org} stocks under the tree species due to fertilization is 2980 kg ha^{-1} (approximately 9%) in the 12 years since establishment of the plantation. On that basis an accumulation rate of 250 kg C ha^{-1} y^{-1} which is linked to fertilization can be calculated (Table 8). These results are comparable with those of Makeschin (1994), who found an increase in soil C_{org} of about 10% after 9 years in the upper 10 cm under a poplar plantation. An ecosystem simulation model developed by Seely et al. (2002) also detected a slight increase in soil organic carbon under aspen after 30 years.

The higher C_{org} stocks under the fertilized block A of annual crops by comparison with block D can be attributed to increased yields due to fertilization, which are accompanied by higher organic matter input. The observed difference between the soil carbon pools under the different fertilized blocks of the two tree species cannot be explained this way, as fertilization hardly increased the yield of willows and poplars. By contrast growth of the undersown grass on the willow plot and/or wild growing weeds under poplars and willows was enhanced remarkably on the fertilized plots. Hence grass and weeds may be responsible for higher carbon stocks on fertilized sites - not trees.

Furthermore, the effects of nitrogen fertilization on decomposition processes must be considered. Referring to the results of Neff et al. (2002) it must be expected that the decomposition of light soil carbon fractions was increased on the A blocks, while soil carbon compounds in mineral associated fractions were stabilized. As soil carbon was not analyzed according to its fractions, the two contrary effects could not be quantified, however.

CONCLUSION

Ligno-cellulosic energy crops may be produced in an environment-compatible and energy-efficient manner. Appropriate species make do with few or no fertilizers and pesticides, cause few emissions during growth and combustion, and enhance the carbon sequestration in soil.

Out of the nine energy crop species investigated, the perennials, especially the woody species such as poplars and willow, proved to be particularly favorable. They provide acceptable yields of 8 to 10 t$_{DM}$ ha^{-1} y^{-1}, which are in the range of established haulm-type crops on sandy soils in Central Europe. On arable farm land, and at least over a period of 14 years, they achieve these yields even without any fertilization. Only willow seems to need a small nitrogen application rate during the first few years. However, conventional haulm-type crops such as whole crop cereals, hemp, topinmabur and grass have an annual yield loss of 0.5 to 2% on low-fertilized fields (75 kg N ha^{-1} and biomass ash) and a loss of 2 to 7% on non-fertilized sites, compared with a conventional application rate of 150 kg N ha^{-1} with traditional basic fertilization.

With 0.2 to 1.2% nitrogen, poplars and willows have only half the N content of whole crop cereals and hemp, and approximately only one third that of cocksfoot grass. Only topinambur has similarly low contents. Due to this low N content, these species cause low NO_x emissions during combustion.

Nitrogen oxides are emitted not only during energy conversion processes but also during the production of energy crops. N_2O flux measurements, carried out over nine years, show that N fertilization level as well as crop species, weather conditions and other variables influence the N_2O emission budget.

Regarding the crop species, it can be summarized that short rotation coppices cause on average lower N_2O emissions than cereals and grass, independently of the fertilization. For instance, poplar and willow fields (without hot spots) emit only 50 to 80% of the N_2O emitted by winter rye and triticale fields. Compared with the conventionally fertilized variants of these cereals, the unfertilized poplars and willows actually cause only 20 to 60% of the N_2O emissions. Only hemp seems to be more favorable. However, this species was only measured in 2000 and 2005, years with low and moderate N_2O emissions, respectively.

N_2O measurements on soils need long periods because periodical and stochastic variations are evident. Whereas in the course of autumn and winter the emissions were comparatively low (below 30 µg N_2O m^{-2} h^{-1}), the N_2O fluxes had peaks above 100 µg N_2O m^{-2} h^{-1} after nitrogen fertilization in spring. In summer and autumn, increased N_2O flux rates were observed from several fertilized blocks. N_2O hot spots (long-lasting N_2O emission spots with more than 200 µg N_2O m^{-2} h^{-1}) can produce up to 3 g N_2O m^{-2} y^{-1} (or in other units: 20 kg N_2O-N ha^{-1} y^{-1}) in the course of few months. Thus they can cause a local measured emission factor of more than 10 % at the corresponding measuring block. As N_2O hot spots were found on a few fertilized blocks only, they do not cause drastic changes in the nitrogen conversion factors if the mean is taken over a greater number of measurement spots. Therefore the mean values of all measurements (all blocks and all years) indicate that the difference in the nitrogen conversion factor with and without N_2O hotspots is only 0.1%.

The mean nitrogen conversion factor, which describes the N_2O greenhouse burden of fertilization, is about 0.8±0.1 % and thus slightly lower than the default value of 1% for N_2O inventories. Therefore, the production of lingo-cellulosic energy crop species under the conditions considered will not lose its CO_2 advantage by nitrogen fertilizing as long as fertilizing results in an adequately higher biomass yield.

Another favorable CO_2-related aspect of the production of energy crops is the long-term sequestration of carbon in soil (C_{org}). However, there are relevant differences between the crop species and the fertilization levels. With approximately 1.3 t ha^{-1} y^{-1} poplars and willows cause significantly higher C sequestration rates in soil than annual crops such as rye and triticale. The non-fertilization on poplar and willow fields causes only a slight reduction of the C_{org} sequestration rate of less than 0.35 t ha^{-1} y^{-1} (probably caused by the lower growth of weeds), whereas the C_{org} loss on non-fertilized fields of annual haulm-type crops is about 0,53 t ha^{-1} y^{-1} related to conventionally fertilized sites. The reason for this fertilization effect on the carbon sequestration under haulm-type crops is supposed to lie in the higher organic matter input due to higher yields.

The most important fact for production of energy crops is the energy efficiency. All of the crop species investigated have excellent energetic input/output ratios between 1:6 and 1:70. However, more important is the gain of energy per unit area resulting from the difference between the energy yield and the cumulated energy demand for production and harvesting. With the exception of cocksfoot grass, topinambur haulm and the poplar variety NE 42, this varies approximately between 100 and 160 GJ ha^{-1} y^{-1}, i. e. 2,600 to 4,300 liter oil equivalents per annum, and depends on the species, the production technology and the fertilization level. Poplar Japan 105 without undersown grass has the highest energy gain, followed by hemp and whole crop cereals. Very interesting is the fact that reduced fertilization results in only a marginal loss of energy gain and non-fertilization of poplar even brings an increase. This means that it is possible to produce lingo-cellulosic energy crops, with lower fertilization rates than conventional food crops which - together with the low pesticide demand - is a further relevant environmental and energetic advantage.

REFERENCES

Anonymous. *Betriebsplanung Landwirtschaft 2004/2005 - KTBL-Database.* KTBL-Schriften-Vertrieb. Landwirtschaftsverlag Münster: Münster, 2004.

Anonymous. *Agrarbericht 2007 zur Land- und Ernährungswirtschaft des Landes Brandenburg.* MLUV Potsdam, 2007.

Anonymous. *Zahlen und Fakten - Energiedaten Nationale und Internationale Entwicklung.* BMWi Berlin, 2008.

Augustin, J.; Merbach, W.; Steffens, L.; Snelinski, B. Nitrous oxide fluxes of disturbed minerotrophic peatlands. *Agribiological Research.* 1998, 51, 47-57.

Bassam, El N. *Energy Plant Species.* James & James Ltd London. 1998.

Bouwman, A.F. Exchange of greenhouse gases between terrestrial ecosystems and the atmosphere. In: Bouwman A.F. (ed.): *Soils and the Greenhouse Effect.* John Wiley & Sons. Chichester New York: 1990, 61-127.

Bouwman, A.F. Direct emission of nitrous oxide from agricultural soils. Nutr. *Cycl. Agroecosys.* 1996, 46, 53-70.

Bouwman, A.F.; Boumans, L.J.M.; Batjes, N.H. Emissions of N_2O and NO from fertilised fields: Summary of available measurement data. *Glob. Biogeochem. Cycles.* 2002, 16, 1058.

Bungart, R. Erzeugung von Biomasse zur energetischen Nutzung durch den Anbau schnellwachsender Baumarten auf Kippsubstraten des Lausitzer Braunkohlereviers unter besonderer Berücksichtigung der Nährelementversorgung und des Wasserhaushaltes. *Dissertation,* Brandenburg. Techn. Universität Cottbus, 1999.

Christensen, S.; Simkins, S.; Tiedje, J.M. Spatial Variation in Denitrification: Dependency of Activity Centers on the Soil Environment. *Soil Science Society of America Journal.* 1990, 54, 1608-1613.

Crutzen, P. J.; Mosier, A.R.; Smith, K. A.; Winiwarter, W. N_2O release from agro-biofuel production negates global warming reduction by replacing fossil fuels. *Atmos. Chem. Phys. Discuss.* 2007, 7, 11191–11205. Available at: http://www.atmos-chem-phys-discuss.net/7/11191/2007/acpd-7-11191-2007.pdf (verified February 24, 2008).

De Klein, C.; Novoa, R.S.A.; Ogle, S.; Smith, K. A.; Rochette, P.; Wirth, T. C.; McConkey, B. G.;. Walsh, M.; Mosier, A.; Rypdal, K.; Williams, S. A. (2006). N_2O Emissions from Managed Soils. and CO_2 Emissions from Lime and Urea Application. In Eggleston, H.S.; Buendia, L.; Miwa, K.; Ngara, T.; Tanabe, K. (eds.) (2006) IPCC Guidelines for National Greenhouse Gas Inventories. National Greenhouse Gas Inventories Programme. *IGES.* Japan: 11.11. Available at: http://www.ipcc-nggip.iges.or.jp/public/2006gl/pdf/ 4_Volume4/V4_11_Ch11_N2O&CO2.pdf (verified February 24, 2008).

Dobbie, K.E.; Smith, K.A. Nitrous oxide emission factors for agricultural soils in Great Britain: The impact of soil water-filled pore space and other controlling variables. *Global Change Biology.* 2003, 9, 204–218.

Fahrmeir, L.; Künstler, R.; Pigeot, I.; Tutz, G. *Statistik – Der Weg zur Datenanalyse.* Berlin, Heidelberg, New York. Springer Verlag. 2001.

Firestone, M.K. Biological denitrification. In: Stevenson, F.J. (ed.): *Nitrogen in Agricultural Soils. American Society of Agronomy.* Madison: 1982, 289–326.

Flessa, H.; Wild, U.; Klemisch, M.; Pfadenhauer, J. Nitrous oxide and methane fluxes from organic soils under agriculture. *European Journal of Soil Science.* 1998, 49, 327-335.

Freney, J. R. Emission of nitrous oxide from soils used for agriculture. *Nutr. Cycl. Agroecosys.:* 1997, 49, 1–6.

Frolking, S.E.; Mosier, A.R.; Ojima, D.S.; Li, C.; Parton, W.J.; C.S. Potter, E.; Priesack, R.; Stenger, C.; Haberbosch, P.; Dorsch. Flessa, H.; Smith, K.A. Comparison of N_2O emissions from soils at three temperate agricultural sites: simulations of year-round measurements by four models. *Nutr. Cycl. Agroecosys.* 1998, 52, 77-105.

Granli, T.; Bøckman, O.C. Nitrous oxide from agriculture. *Norwegian J. Agric. Sci. Suppl.* 1994, 12, 1-128.

Grigal, D. F.; Berguson, D. W. Soil carbon changes associated with short-rotation systems. *Biomass and Bioenergy.* 1998, Vol. 14(4), 371 - 377.

Hartmann, H.; Schmid, V. *NOx Emissionen in einer Hackschnitzelfeuerung.* Note, Landtechnik Weihenstephan 2001.

Hellebrand, H.J.; Kern, J.; Scholz, V. Long-term studies on greenhouse gas fluxes during cultivation of energy crops on sandy soils. *Atmos. Environ.* 2003, 37, 1635-1644.

Hellebrand, H.J.; Scholz, V.; Kern, J.; Kavdir, Y. N_2O release during cultivation of energy crops. *Agrartechnische Forschung.* 2005, 11, E114-E124.

IPCC (2007): IPCC WG1 AR4 Report. Technical Summary. Table TS2. In: *Climate Change 2007 - The Physical Science Basis Contribution of Working Group 1 to the Fourth Assessment Report of the IPCC. Cambridge University Press:* 33. Available at: http://ipcc-wg1.ucar.edu/wg1/Report/AR4WG1_Print_TS.pdf (verified February 24, 2008).

Jackson, L.E.; Calderon, F.J.; Steenwerth, K.L.; Scow, K.M.; Rolston, D.E. Responses of soil microbial processes and community structure to tillage events and implications for soil quality. *Geoderma.* 2003, 114, 305–317.

Jug, A. Standortkundliche Untersuchungen auf Schnellwuchsplantagen unter besonderer Berücksichtigung des Stickstoffhaushalts, *Dissertation,* LMU Munich, 1998.

Kahle, P.; Boelcke, B. Auswirkungen des Anbaus schnellwachsender Baumarten im Kurzumtrieb auf ausgewählte Bodeneigenschaften. *Research Report of ATB.* Vol. 35. Potsdam-Bornim. 2004.

Kaiser, E.A.; Kohrs, K.; Kücke, M.; Schnug, E.; Heinemeyer, O.; Munch, J.C. Nitrous oxide release from arable soil: importance of N fertilization, crops and temporal variation. *Soil Biology & Biochemistry.* 1998, 30, 1553–1563.

Kaiser, E.A.; Ruser, R. Nitrous oxide emissions from arable soils in Germany – An evaluation of six long-term field experiments. *J. Plant Nutr. Soil Sci.* 2000, 163, 249-260.

Karpenstein-Machan, M. *Auswirkungen von pestizidfreiem Energiepflanzenanbau auf die Biomasseerträge. energie pflanzen.* 2000, Vol. II, 36-38.

Lehn, J.; Wegmann, H. *Einführung in die Statistik.* Wiesbaden, Teubner Verlag. 2006.

Loftfield, N.; Flessa, H.; Augustin, J.; Beese, F. (1997): Automated gas chromatographic system for rapid analysis of the atmospheric trace gases methane, carbon dioxide, and nitrous oxide. *J. Environ. Qual.* 26: 560-564. and "Loftfields Analytical Solutions". Dr. Norman, S. Loftfield GbR. http://www.loftfield.de/index_e.htm (verified February 24, 2008).

Makeschin, F. Effects of energy forestry on soils. *Biomass and Bioenergy.* 1994, Vol. 6 (1/2), 63-79.

Mogge, B.; Heinemeyer, O.; Kaiser, E.-A.; Munch, J.C. N_2O-emissions of forest soils in Northern Germany - Seasonal variability and influencing parameters. In: Van Cleemput, O.; G. Hofman. and A. Vermoesen (eds.): *Progress in nitrogen cycling studies. Kluwer Academic Publishers.* Dordrecht: 1996, 585-588.

Mulvaney, R.L.; Khan, S.A.; Mulvaney, C.S. Nitrogen fertilizers promote denitrification. *Bio. Fertil. Soils.* 1997, 24, 211-220.

Neff, J. C. et al. Variable effects of nitrogen additions on the stability and turnover of soil carbon. 2002, *Nature* Vol. 419, (letters to Nature), 915-917.

Novoa, R.S.A.; Tejeda, H.R. Evaluation of the N_2O emissions from N in plant residues as affected by environmental and management factors. *Nutr. Cycl. Agroecosys.* 2006, 75, 29–46.

Nussbaumer, T. Primär- und Sekundärmaßnahmen zur NOx-Minderung bei Biomassefeuerungen. In: *VDI-Bericht.* Vol. 1319. VDI-Verl. Düsseldorf: 1997, 141-166.

Obernberger, I. Nutzung fester Biomasse in Verbrennungsanlagen unter besonderer Berücksichtigung des Verhaltens aschebildender Elemente. In *Schriftenreihe Thermische Biomassenutzung.* Technische Universität Graz. 1997.

Röver, M.; Heinemeyer, O.; Kaiser, E.-A. Microbial induced nitrous oxide emissions from an arable soil during winter. *Soil Biology & Biochemistry.* 1998, 30, 1859-1865.

Röver, M.; Heinemeyer, O.; Munch, J.C.; Kaiser, E.-A. Spatial heterogeneity within the plough layer: high variability of N2O emission rates. *Soil Biology & Biochemistry.* 1999, 31, 167-173.

Ruser, R.; Flessa, H.; Russow, R.; Schmidt, G.; Buegger, F.; Munch, J.C. Emission of N_2O. N_2 and CO_2 from soil fertilized with nitrate: effect of compaction. soil moisture and rewetting. *Soil Biology & Biochemistry.* 2006, 38, 263-274.

Scheffer, F.; Schachtschabel, P. *Lehrbuch der Bodenkunde.* Heidelberg. Spektrum Akademischer Verlag GmbH. 2002.

Scholz, V.; Hellebrand, H.J., Grundmann, P. *Produktion von nachwachsenden Energierohstoffen auf landwirtschaftlichen Flächen.* KTBL-Schrift 420, Osnabrück, Landwirtschaftsverlag Münster, 2004,167-181.

Scholz, V. Methodik zur Ermittlung des Energieaufwandes pflanzenbaulicher Produkte am Beispiel von Biofestbrennstoffen. *Agrartechnische Forschung.* 1997, Vol. 1, 11-18.

Scholz, V.; Berg. W.; Kaulfuß, P. Energy Balance of Solid Biofuels. *J. agric. Engng Res.* 1998, Vol. 71, 263-272.

Scholz, V.; Ellerbrock, R. The growth productivity and environmental impact of the cultivation of energy crops on sandy soil in Germany. *Biomass and Bioenergy.* 2002, Vol. 23, 81-92.

Scholz, V.; Kaulfuss, P. Energiebilanz für Biofestbrennstoffe. *Research report of ATB.* Potsdam-Bornim. 1995, Vol. 3.

Scholz, V.; Krüger, K.; Höhn, A. et al. *Umwelt- und technologiegerechter Anbau von Energiepflanzen. Research report of ATB.* Potsdam-Bornim, 1999, Vol. 1.

Seely, B. et al. Carbon sequestration in a boreal forest ecosystem: results from the ecosystem simulation model. FORECAST. *Forest Ecology and Management.* 2002, Vol. 169, 123-135.

Stehfest, E.; Bouwman, L. N_2O and NO emission from agricultural fields and soils under natural vegetation: summarizing available measurement data and modelling of global annual emissions. *Nutr. Cycl. Agroecosyst.* 2006, 74, 207-228.

Stetter, U.; Makeschin, F. *Modellvorhaben "Schnellwachsende Baumarten": Humusgehalt ehemals landwirtschaftlich genutzter Böden nach Aufforstung mit schnellwachsenden Baumarten.* Bonn, 1999.

Teepe, R.; Brumme, R.; Beese, F. Nitrous oxide emissions from soil during freezing and thawing periods. *Soil Biology & Biochemistry.* 2001, 33, 1269–1275.

Teepe, R.: Vor, A.; Beese, F.; Ludwig, B. Emissions of N_2O from soils during cycles of freezing and thawing and the effects of soil water, texture and duration of freezing. *European Journal of Soil Science.* 2004, 55, 357–365.

Ulzen-Appiah, F. et al. Soil carbon pools in short rotation willow (Salix dasyclados) plantation four years after establishment. *Bioenergy.* 2000.

Van der Weerden, T.J.; Sherlock, R.R.; Williams, P.H.; Cameron, K.C. Nitrous oxide emissions and methane oxidation by soil following cultivation of two different leguminous pastures. *Biol. Fertil. Soils.* 1999, 30, 52-60.

VDI 4600. *Kumulierter Energieaufwand - Beispiele.* VDI-Richtlinie 4600 Blatt 1, VDI Düsseldorf, 1998, 1-39.

Veldkamp, E.; Keller, M. Nitrogen oxide emissions from a banana plantation in the humid tropics. *J. Geophys. Res.* 1997, 102, 15889–15898.

Wanga, H.; Wanga, W.; Yina, C.; Wang, Y.;. Lua, J. Littoral zones as the "hotspots" of nitrous oxide (N_2O) emission in a hyper-eutrophic lake in China. *Atmos. Environ.* 2006, 40, 5522-5527.

In: New Research on Biofuels
Editors: J. H. Wright and D. A. Evans

ISBN 978-1-60456-828-8
© 2008 Nova Science Publishers, Inc.

Chapter 5

EXERGETIC ANALYSIS OF BIOFUELS PRODUCTION

K. J. Ptasinski

Department of Chemical Engineering, Eindhoven University of Technology,
P.O. Box 513, 5600 MB Eindhoven, the Netherlands

ABSTRACT

Biomass is a key feedstock to produce renewable biofuels such as Fischer-Tropsch hydrocarbons, methanol, and hydrogen. However, limitation of land and water, and competition with food production reduce potential significance of biomass as a renewable energy source. The development of efficient conversion technologies, which are able to compete with fossil fuels, is a key challenge for biomass-based systems.

Production of biofuels is traditionally analyzed by energetic analysis based on the First Law of Thermodynamics. However, this type of analysis shows only the mass and energy flows and does not take into account how the quality of the energy and material streams degrades through the process. In this chapter the exergy analysis, which is based on the Second Law of Thermodynamics, is used to analyze the conversion of biomass to biofuels.

The most promising biomass-to-biofuel route is a two-stage process involving production of syngas from biomass gasification, followed by synthesis of transportation fuels. Several overall technological chains biomass-to-biofuels are evaluated, including Fischer-Tropsch hydrocarbons, hydrogen, and methanol. It is shown that that exergetic efficiency of production of biofuels is lower than that for fossil fuels.

Biomass gasification shows the largest exergy losses in the overall chain biomass-to-biofuels. Modification of process conditions to improve efficiency of gasification is discussed. It is shown that the optimal gasification conditions correspond to the carbon boundary point where all carbon present in biomass is gasified. The efficiency of biomass gasification can be also improved in a thermal pretreatment called torrefaction.

NOMENCLATURE

I   exergy loss (irreversibility)
P   pressure
P_o environmental pressure
Q   heat
T   temperature
T_o environmental temperature
W   work

Greek

ε_{ch} specific chemical exergy
ε_{ph} specific physical exergy
ε_t specific exergy of material stream
E_i exergy rate of material stream
E^Q thermal exergy rate
E^W exergy rate due to work interaction
Π entropy production
τ Carnot efficiency

INTRODUCTION

Two global problems related to the use of fossil fuels are a fast depletion and environmental damage due to emissions of various gases mainly CO_2 [1, 2]. It is commonly believed that the solution to the global problems would be to replace the existing fossil fuels by renewable energy (e.g. solar, wind and biomass). Biomass is considered the renewable energy source with the highest potential to contribute to energy needs of the modern society for both developed and developing economies world-wide. Biomass is a renewable and relatively clean feedstock for producing modern energy carriers such as electricity and transportation fuels [3, 4]. The key features of biomass are renewability and neutral CO_2 impact.

Transportation fuels derived from biomass, such as methanol, Fischer-Tropsch hydrocarbons, and hydrogen, are gaining currently more attention as potential substitute for fossil fuels [5]. The most promising biomass-to-biofuel route is a two-stage process involving production of syngas from biomass gasification, followed by synthesis of transportation fuels. During gasification process biomass is partially oxidized to a syngas, which can be used to produce transportation fuels and chemicals in catalytic reactions, and also for electricity generation in gas turbines, engines, fuel cells.

However, limitation of land and water, and competition with food production reduce potential significance of biomass as a renewable energy source [6]. The average conversion efficiency of sunlight into chemical energy in biomass through photosynthesis is about 0.5-1.0% what is much lower compared to other forms of renewable energy such as photovoltaics or wind energy [7]. In practice various biomass feedstocks can be used to produce biofuels and the question is whether all the biomass types can be converted to biofuels with reasonable conversion efficiency.

Therefore the development of efficient conversion technologies, which are able to compete with fossil fuels, is a key challenge for biomass-based systems. One of the difficulties with measuring energy efficiency is the lack of consensus on the evaluation of performance of different stages of energy systems. In practice various energy performance indicators are used, usually based on thermodynamics or economics. The thermodynamic indicators can evaluate either the first-law or the second-law efficiency based on exergy concept. Second-law based thermodynamic indicators are nowadays commonly accepted as the most natural way to measure the performance of different processes, ranging from energy technology, chemical engineering, transportation, agriculture, etc [8, 9].

The purpose of this chapter is to present an exergy analysis of biomass-to-biofuels chains. The chapter starts with a short introduction on evaluation of exergetic efficiency. In the next section the exergetic efficiency of biomass-to-biofuels processes, including Fischer-Tropsch hydrocarbons, hydrogen and methanol is evaluated. Subsequently, the improvement of exergetic efficiency of biomass gasification, which is the least-efficient unit operation in the whole biomass-to-biofuel chain, is presented.

EVALUATION OF EXERGETIC EFFICIENCY

There are two key aspects in the development of future biomass-to-biofuels systems, namely development of chemical conversion technologies and selection of the most efficient chains: biomass feedstock-conversion process-biofuel. The first aspect relates to solving relevant chemical and chemical engineering problems, such as reduction of tar content in the syngas from biomass gasification [10], design of biomass gasifiers [11], and development of catalysts for chemical conversion of syngas to biofuels [12]. The scientific challenges belonging to these problems show similar character to those encountered for fossil fuel technology.

On the other hand the scientific and engineering challenges resulting from the second development key-aspect are quite unique for biomass. Due to a large variation of biomass feedstocks, conversion technologies, and biofuels, the future bioenergy systems can be designed almost from scratch. To this end the selection of the most efficient biomass conversion processes and synthesis of biofuels requires the correct use of thermodynamics. It is believed that for the successful introduction of biofuels both development aspects, namely chemical and technological developments as well as thermodynamic analysis of biomass-to biofuels processes should go hand-in-hand.

Exergy analysis is a relatively new method of thermodynamic analysis that has recently been applied in different fields of engineering and science [13]. The exergy method takes into account not only the quantity of materials and energy flows, but also their quality as well. The main reason of exergy analysis is to detect and evaluate quantitatively the losses that occur in thermal and chemical processes. Exergy is defined as the maximum amount of work that can be obtained from a material stream, heat stream or work interaction by bringing this stream to environmental conditions [8,14]. The term environment is regarded as a medium composed of common substances existing in abundance within the Earth's atmosphere, oceans, and crust. The reference state is usually taken to be at standard temperature (T_o = 298.15 K) and pressure (p_o = 1 atm). Some reference species include CO_2, O_2, N_2, having a mole fraction of 0.0003, 0.2099, 0.7903 in dry air, respectively.

Among the different forms of exergy, three forms are the major contributors to total exergy and they are: thermal exergy, work exergy, and exergy of material. The exergetic value of a heat flow that is its quality of energy, is the maximum amount of work that could be obtained from it by using the environment as a reservoir of zero-grade thermal energy.

$$E^Q = Q\tau = Q(1 - \frac{T_o}{T}) \tag{1}$$

Work interaction is a completely ordered form of energy and therefore the exergy value E^W equals to the amount of work done.

$$E^W = W \tag{2}$$

The quality of a material stream can be expressed using their physical and chemical exergy.

$$\varepsilon_t = \varepsilon_{ph} + \varepsilon_{ch} \tag{3}$$

The physical exergy ε_{ph} is equal to the maximum amount of work obtainable when a compound or mixture is brought from its temperature T and pressure P to environmental conditions, characterized by environmental temperature T_0 and P_0. The standard chemical exergy of a pure chemical compound ε_{ch} is equal to the maximum amount of work obtainable when a compound is brought from the environmental state, characterized by the environmental temperature T_0 (298.15K) and environmental pressure P_0 (1 atm), to the dead state, characterized by the same environmental conditions of temperature and pressure, but also by the concentration of reference substances in standard environment.

The exergy balance of a process can be represented in the following form using exergy values of all streams entering and leaving the process:

$$\sum_{IN} E_j + E^Q + E^W = \sum_{OUT} E_k + I \tag{4}$$

where $\sum_{IN} E_j$ and $\sum_{OUT} E_k$ are exergy flow of all entering and leaving material streams, respectively, E^Q and E^W are the sums of all thermal exergy and work interactions involved in a process. The difference between the concept of exergy and those of mass and energy is that exergy is not conserved but subjected to dissipation. It means that the exergy leaving any process step will always be less than the exergy in. The difference between all entering exergy streams and that of leaving streams is called irreversibility I. Irreversibility represents the internal exergy loss in process as the loss of quality of materials and energy due to dissipation. Irreversibility relates also to entropy production Π in the system and can be expressed as follows:

$$I = T_o \Pi \tag{5}$$

EXERGETIC EFFICIENCY OF BIOMASS-TO-BIOFUEL PROCESSES

Three primary methods, namely biomass gasification, pyrolysis and liquefaction, and biomass hydrolysis, can be applied to produce biofuels [15]. In this section the production of potential transportation fuels, including Fischer-Tropsch hydrocarbons, hydrogen, and methanol, from various biomass feedstocks is evaluated using exergy analysis. For all these biofuels the most promising route is used involving production of syngas from biomass gasification, followed by synthesis of biofuels [16, 17]. First, the exergy analysis of Biomass Integrated Gasification Fischer-Tropsch (BIG FT) process is presented. This process combines an air-blown gasifier using sawdust as a feedstock, with a Fischer-Tropsch reactor and steam-Rankine cycle for electricity generation from the FT tail gas. Next, the exergy analysis of hydrogen production by gasification of various biomass feedstocks, such as wood, vegetable oil, and manure, is discussed. Subsequently, a new process to make methanol from sewage sludge is analyzed. Finally, the efficiency of production of the above-mentioned biofuels is compared, using the efficiency of hydrogen-from-natural gas process as a reference.

Figure 1 shows a block diagram of conversion process for production of biofules from biomass gasification. A wide range of biomass sources, such as traditional agricultural crops, residues from agriculture and foresting can be used to make biofuels [18]. These biomass feedstocks vary greatly in chemical composition, energy content, ash and moisture content. This is generally regarded as a real advantage, because it means that the best and economically attractive feedstock can be selected. For wet biomass feedstocks pretreatment, in particular drying, is needed to adjust the moisture content into a desired value.

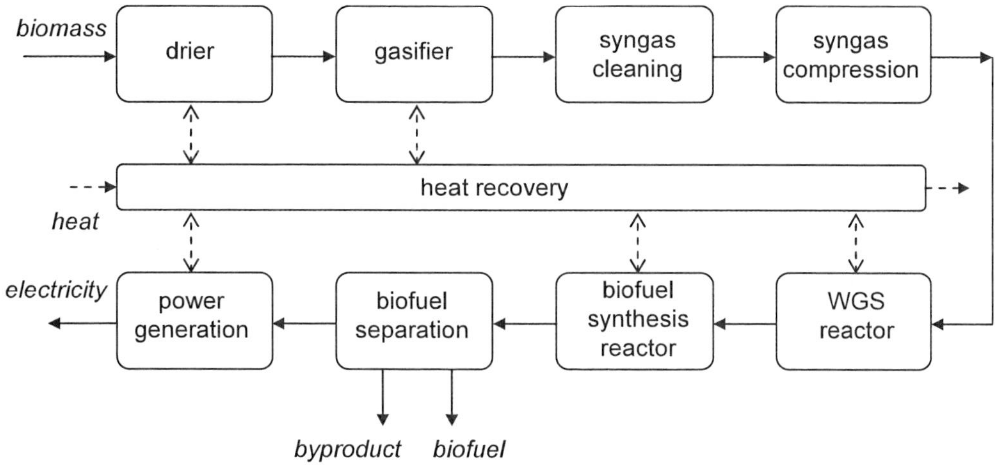

Figure 1. Schematic flowsheet of the biomass-to-biofuel conversion process.

In the gasifier biomass reacts at high temperature (typically 900°C) with air, oxygen, or/and steam to produce a gas product called syngas of producer gas. The main gas components present in syngas are CO, H_2, CO_2, CH_4, H_2O, and N_2. Syngas components carbon monoxide and hydrogen are the building reactants to synthesize different biofuels. In the gasifier some undesired components, such as tars, H_2S, NH_3, and HCl, are also produced, which are harmful for the biofuel synthesis. The purpose of gas cleaning is to remove all toxic and harmful components either present in biomass or formed during gasification. Syngas cleaning can involve high temperature catalytic tar removal, wet scrubbers to remove gas impurities and solid particles, and fabric filters for removal of fine dust.

The synthesis of most biofuels from syngas is usually a high pressure process and therefore syngas after cleaning unit is compressed up to the desired reactor pressure. Before the biofuel synthesis the ratio between CO and H_2 in syngas is adjusted in a water gas shift (WGS) reactor, according to the following catalytic chemical reaction:

$$CO + H_2O = H_2 + CO_2 \tag{6}$$

Subsequently, syngas reacts in a catalytic synthesis reactor where specific biofuels, such as Fischer-Tropsch hydrocarbons or methanol are produced. In case of hydrogen production, only WGS reactors are sufficient and no specific synthesis reactors are needed. The biofuel formed in the synthesis reactor is separated from by-products in a separation unit. The remaining by-products contain usually high energy content and they can be used to produce electricity in a power generation unit or process steam in the heat integration unit. The heat recovery is particularly needed as in the overall process synthetic gas is cooled from a high temperature in the gasifier to much lower temperature in the biofuel synthesis reactor.

Fischer-Tropsch Hydrocarbons from Wood

Currently, various biomass-to-liquid (BTL) processes are under development and the most promising is a two-stage process where biomass is first gasified, followed by conversion of the produced synthesis gas to Fischer-Tropsch (FT) hydrocarbons [19]. Air can be used as gasification medium in order to avoid an oxygen separation process, which is expensive on the relatively small scale of most biomass gasifiers. The Fischer-Tropsch process has a long history of about seventy years [20]. The overall chemical reaction of converting syngas into hydrocarbons is:

$$n\, CO + 2n\, H_2 = [-CH_2-]_n + n\, H_2O \tag{7}$$

The Fischer-Tropsch process based on sawdust as a feedstock is recently developed according to the schematic flowsheet shown in Fig.1, and evaluated using exergy analysis [21]. This feedstock contains 9.3% fixed carbon, 55% volatile matter, 0.7 % ash, and 35.0% moisture. The higher heating value of sawdust is 12.63 MJ/kg. The biomass is dried to 10 wt% moisture by indirect drying using reaction heat from the Fischer-Tropsch section. Dried biomass is autothermally gasified with air at a temperature of 900°C and atmospheric pressure. The product gas is cooled to 90°C, generating 50 bar and 20 bar steam, which are used in steam cycles for electricity production. The gas is subsequently cleaned from ash, ammonia and salts, and subsequently compressed to 25 bar. The raw synthesis gas from biomass gasifier has relatively low H_2/CO ratio which is increased in a water-gas-shift (WGS) catalytic reactor [22] due to chemical reaction Eq. 6. The feed gas to the FT reactor has a typical composition of approximately 26% H_2, 12% CO, 17% CO_2, 44% N_2, and small amounts of CH_4 (dry basis). The gas is converted to gaseous and liquid hydrocarbons at a temperature of 260°C by cobalt-catalyzed Fischer-Tropsch synthesis. The total single-pass (H_2+CO) conversion is 80%. The products from the FT reactor are cooled to 40°C, so that hydrocarbon liquids are condensed from the tail gas, which contains most of the naphtha (C5-C8). Diesel (C9-C22) and wax (C23+) are recovered as liquid products. The tail gas is incinerated in a boiler and electricity is generated in a traditional steam-Rankine cycle by means of a condensing steam turbine.

In this process 57 ton/h of biomass is converted into 5.6 ton/h of liquid FT products (diesel (C9-C22) and wax (C23+)) and 4.0 MW of electricity is also net produced. On weight basis the net output of hydrocarbon fuel is only 15.2%. The total amount of produced electricity is 25.1 MW but 20.1 MW of electricity is used in the plant for internal purposes. The exergy analysis shows that the biomass feed contains 210.8 MW exergy and the final liquid products, diesel (C9-C22) and wax (C23+), contain 72.9 MW with co-production of 4.0 MW of electricity. The detailed mass, energy and exergy balances for this process are available elsewhere [21]. Table 1 summarizes the main components of exergy balance, normalized to 100% of biomass exergy as the main exergetic input to this process. The only input to the process is the exergy of biomass as all process utilities are produced internally. The exergy output of the process consists of 72.9 MW of FT liquid, that corresponds to 34.6% of biomass exergy, and 4.0 MW of electricity produced net in this process (1.9% of biomass exergy). The overall exergy losses are the difference between the exergy input and exergy output and are equal to 63.5% of biomass exergy. The overall exergetic efficiency of this process is 36.4% calculated as the ratio of exergy output and input to the process.

The largest exergy losses in this process occur in biomass gasification and generation of power from FT tail gas, as shown in Table 2. The high exergy losses in the gasifier are mainly due to large entropy generation in irreversible chemical reactions taking place in the gasifier where wood is converted into syngas.

Table 1. Main components of exergy balance and exergetic efficiencies for biomass-to-biofuels plants (in %)

Process	Exergy input		Exergy output		Exergy losses		Exergetic efficiency
	Biomass feed	Process utilities	Fuel product	Process utilities	Overall	Gasifier	
FT fuel from wood	100	0	34.6	1.9	63.5	21.7	36.4
Hydrogen from wood	100	24.0	55.0	26.8	42.3	28.3	65.7
Methanol from sewage sludge	100	74.1	69.7	28.9	75.4	27.2	56.0

Table 2. Relative exergy losses (%) per process unit for biomass-to-biofuel plants

Process unit	FT fuel from wood	Hydrogen from wood	Methanol from sludge
Biomass drying	4.3	0.4	18.2
Biomass gasification	34.2	66.9	36.1
Syngas cleaning	0.6	0.8	1.5
Syngas compression	4.8	6.3	28.5
WGS reactors	0.8	8.2	-
Biofuel synthesis reactor	2.9	-	2.2
Biofuel separation	2.0	10.2	1.5
Heat recovery	14.2	7.2	12.0
Power generation	36.2	-	-

Hydrogen from Biomass Gasification

Hydrogen is one of the most promising energy carriers for the future. The concept of hydrogen economy assumes only two energy carriers: hydrogen and electricity [23, 24]. Currently 95% of hydrogen comes from fossil fuels but biomass has the potential to become a sustainable source of hydrogen [25]. Biomass-to-H_2 conversion technologies can be divided into two categories: thermochemical and biochemical.

The most widely practiced thermochemical process route for biomass-to-hydrogen is gasification coupled with water gas shift, according to the flowsheet shown in Fig. 1. The feed is wood which contains 78.4 wt% organic fraction, 19.8 wt% moisture and 1.84 wt% ash [26]. The wood is first dried in the thermal drier to its final moisture content of 10 wt%. The partially dried wood enters the gasifier, where a syngas containing H_2, CO, CO_2 and H_2O is produced at the temperature of 900°C. The syngas leaving the gasifier is cooled down, cleaned, and compressed to 30 bar. In two water gas shift reactors (400°C, and 150°C, respectively) CO is subsequently converted into H_2 due to reaction Eq. 6. Water present in the gas leaving the shift reactors is separated in a flash unit at 20 bar and 25°C whereas CO_2 is separated from H_2 by pressure swing adsorption (PSA). Separation efficiency of H_2 is 85% and 100% purity of hydrogen is produced.

This plant converts 1000 kg/h wood into 76 kg/h hydrogen and 1330 kg/h CO_2-rich gas as the by-product. The net output of hydrogen on mass basis is thus 7.6% only. The exergy analysis indicates that 4.64 MW of wood exergy is transformed into 2.54 MW exergy of hydrogen and 0.61 MW of CO_2-rich gas.

The main components of exergy balance normalized to 100% of biomass exergy as the main exergetic input to this process are summarized in Table 1. The additional exergetic input to the process (24.0% of biomass exergy) consists of process utilities, mainly oxygen used in the biomass gasifier, steam used in biomass dryer and steam used as a reactant in water-gas-shift reactors, and electricity in syngas compressor. The exergy output of the process consists of hydrogen (55.0% of biomass exergy) and utilities produced in this process (26.6% of biomass exergy), mainly exhaust steam from biomass dryer, steam produced from syngas cooling after gasifier and WGS reactors, as well as exergy of CO_2-rich gas. The overall exergy losses are the difference between the exergy input and exergy output and are equal to

42.3% of biomass exergy. The overall exergetic efficiency of this process is 65.7% calculated as the ratio of exergy output and input to the process.

Exergy analysis of thermochemical H_2 production has been performed for other biomass feedstocks, including vegetable oil and manure, which contain various moisture content: 0, and 43.6 wt%, respectively. The results reveal that the exergetic efficiency decreases with increasing moisture content in the feedstock, as shown in the final section.

Methanol from Sewage Sludge

Methanol may play a significant role as a synthetic fuel for the future as it has higher energy content per volume than the other alternative fuels Moreover, minimal changes in the existing fuel distribution network are required for methanol. Finally, methanol can considerably reduce automotive emissions and requires no antiknock alternatives because of its high octane number.

Generally, methanol can be produced from any organic source including biomass, and municipal wastes [27] according to the flowsheet shown in Fig. 1. Sewage sludge is a by-product of waste water purification and it contains a reasonably high fraction of organic material, which is rich in carbon - the main constituent of methanol [28]. The second constituent of methanol, hydrogen, can be obtained from water present in wet sludge during the gasification process. The sludge from wastewater treatment contains only 1-2 wt% solids and is mechanically (up to 20 wt%) and thermally dewatered. The solids contain 56% of organic materials which are rich in carbon (50.4%), oxygen (30.7%), and hydrogen (6.9%). In the gasifier a synthetic gas is produced at temperature range of 800-1000°C, containing mainly CO and H_2. Next the syngas is cleaned from solid particles and gaseous components, and compressed to a pressure of 77 bar. The methanol synthesis is based on the ICI low pressure methanol process at temperature of 200°C according to chemical reactions:

$$CO + 2H_2 = CH_3OH \tag{7}$$

$$CO_2 + 3H_2 = CH_3OH + H_2 \tag{8}$$

The last step is the methanol separation from dissolved gases and water in a distillation column. A part of the recycle gas from the methanol synthesis and gas from the methanol separation is purged and combusted in a heat recovery unit.

The plant converts 50 000 tons dry solids/year into 4300 tons methanol/year at gasifier temperature of 1000°C and a dry solid content of the sludge leaving the dryer of 80 wt%. The net output of methanol is 8.6% on weight basis but on the exergy basis 22.8 MW of sludge exergy is converted into 15.4 MW of methanol exergy.

Table 1 shows the main components of exergy balance normalized to 100% of biomass exergy as the main exergetic input to this process. The additional exergetic input to the process (74.1% of biomass exergy) consists of process utilities, mainly steam used in dryer and reboiler of distillation column, electricity in both compressors, and additional fuel in biomass gasifier. The exergy of utilities used in this process is relatively high due to a very wet biomass and the overall process in not autotherm. The exergy output of the process consists of methanol (69.7% of biomass exergy) and utilities produced in this process (28.9%

of biomass exergy), mainly heat from methanol reactor, methanol condenser, purge gas burner, and exhaust steam from biomass dryer. The overall exergy losses are the difference between the exergy input and exergy output and are equal to 75.4% of biomass exergy. The overall exergetic efficiency of this process is 56% calculated as the ratio of exergy output and input to the process. A breakdown by plant units of the total exergy loss is shown in Table 2. Most of the exergy losses are associated with gasification, compression and thermal drying.

Comparison of Exergetic Efficiency for Various Biofules

Table 3 summarizes exergetic efficiency of biomass-to-biofuel routes considered in this section. Moreover, the exergetic efficiency of hydrogen production from biomass using biochemical processes such as fermentation and anaerobic digestion is also included [26]. The exergetic efficiency for hydrogen-from-natural-gas process is also reported in this table, as a typical value for fossil fuels [29, 30]. All reported biomass routes have lower exergetic efficiency compared to fossil fuels. The highest efficiency is for hydrogen-from-biomass processes from high quality feedstocks, such as wood or vegetable oil. The latter feedstock, which is carbon-rich and moisture-free can compete with fossil fuels.

The conversion efficiency of all investigated biomass-to-biofuel routes can be increased by improving the operation of biomass gasifier, which shows the highest exergy losses in all considered processes.

IMPROVEMENT OF EXERGETIC EFFICIENCY OF BIOMASS GASIFICATION

In the previous section it was demonstrated that the gasifier is one of the least-efficient unit operation in the production of biofuels from biomass. Therefore the analysis of exergetic efficiency of the gasifier alone can contribute to the efficiency improvement of the whole biomass-to-biofuel chain. In the first part of this section the optimal gasification condition are discussed. Subsequently, the effect of biomass composition on the exergetic efficiency is studied. The final section is devoted to improvement of biomass gasification using a thermal pretreatment called torrefaction.

Table 3. Exergetic efficiency for biomass-to-biofuel processes

Process	Biomass feed	Biofuel	Exergetic efficiency (%)
Fischer-Tropsch	sawdust	biodiesel	36.4
Gasification	wood	hydrogen	65.7
Gasification	vegetable oil	hydrogen	79.1
Gasification	manure	hydrogen	35.8
Methanol-from sludge	sewage sludge	methanol	56.0
Fermentation	potato peels	hydrogen	29.1
Anaerobic digestion	household waste	hydrogen	36.3
SMR	natural gas	hydrogen	78.0

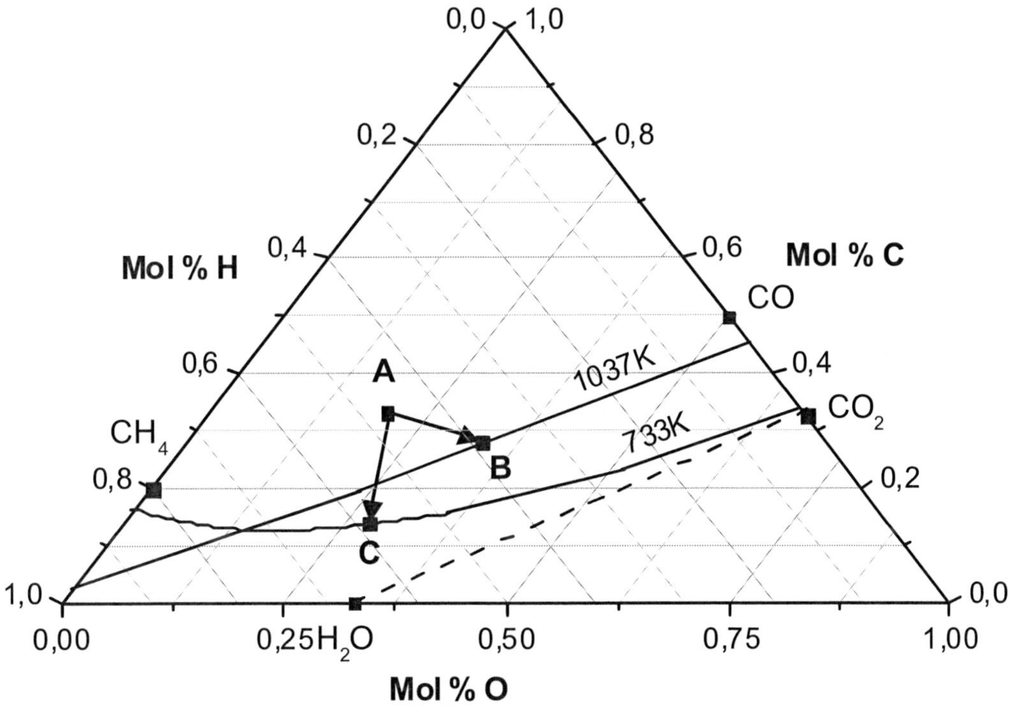

Figure 2. Molar triangular diagram indicating A: biomass feed, B: biomass in equilibrium with air at carbon boundary, C: biomass in equilibrium with steam of 500 K at carbon boundary.

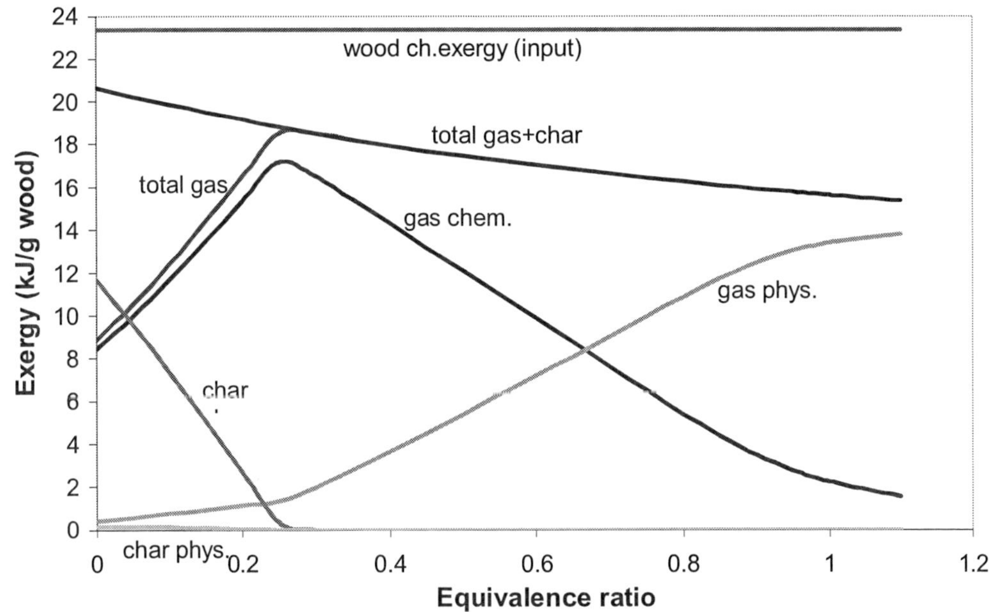

Figure 3. Distribution of exergy in the product gas and char for biomass gasification by air.

Optimal Gasification Conditions for Biomass

The triangular C-H-O diagram, shown in Figure 2, can be used to analyze the efficiency of biomass gasification [31, 32]. The chemical composition at any point in the diagram can be computed at a given temperature and pressure by calculation of the gas phase compositions in equilibrium with solid carbon (graphite) [33]. In the triangular diagram, several lines are shown, the so-called solid carbon deposition boundaries. Above the solid carbon boundary, solid carbon exists in heterogeneous equilibrium with gaseous components, while below the carbon boundary no solid carbon is present.

In the triangular diagram point A represents biomass fuel according to a general formula of $CH_{1.4}O_{0.59}N_{0.0017}$. Adding oxygen to the biomass results in a movement of overall composition to the point B. At this point, all carbon is present in the gaseous phase as carbon monoxide, carbon dioxide or methane. If more oxygen is added, the line between CO_2 and H_2O is crossed, meaning that the fuel has been completely combusted. Similarly, adding steam to the biomass results in a movement of the overall composition to point C, and this way the carbon boundary is also crossed.

Figure 3 shows the distribution of exergy in the syngas during biomass gasification by air for various equivalence ratios [32]. Biomass is gasified using air (both at the temperature of 25°C) in the adiabatic gasifier at atmospheric pressure. Both reagents are brought to chemical equilibrium and the products, gases and char, leave the system at temperature T. The equivalence ratio is the amount of air added to the biomass relative to the amount of air needed for stoichiometric combustion. The chemical exergy of wood represents the only exergetic input for the process as air is introduced at the environmental condition and its exergy is zero. Gas and (unreacted) char are the exergy output from the gasification. The chemical and physical exergy of gas initially increase when air is added due to conversion of solid carbon. At the equivalence ratio of 0.26 the chemical and total exergy of gas reach the maximum at the carbon boundary. Beyond this point the gas composition changes in this way that carbon present as CO is oxidized to form CO_2 and hydrogen present as H_2 is oxidized into H_2O. Due to these exothermic reactions the gas temperature and its physical exergy increase. However, this increase is not compensated by the decrease of gas chemical exergy and therefore the total gas exergy decreases.

The carbon boundary point, which is a sharp maximum, is the optimum point for operating air-blown gasifier. At this point the exergetic efficiency of biomass gasification which is the ratio of the exergetic output (gas and char) to exergetic input (biomass and air) is equal to 80.5%. Gasification at the carbon boundary point is more efficient than slow pyrolysis without addition of air, which has an efficiency of 76.8%. This can be explained by the fact that the exothermic oxidation of carbon by air is coupled to endothermic water-gas reaction and Boudouard reaction. As long as solid carbon is present in the system, addition of air increases the exergy contained in the produced gas.

Influence of Biomass Chemical Composition

Biomass fuels vary in many properties, such as their heating values, proximate analyses (fixed carbon, volatile material, ash content and moisture content), ultimate analyses (amounts of carbon, hydrogen, oxygen, sulphur, nitrogen, chloride and other impurities) and sulphur content. The most important elements in biofuels are carbon, oxygen, and hydrogen, and their relative contents as atomic ratios O/C and H/C are shown in Figure 4 (van Krevelen diagram) for typical biofuels, ranging from coal to biomass [34, 35]. It was demonstrated that O/C ratio has much more influence on biomass properties and gasification performance compared to H/C ratio, e.g. lower heating value of biomass decreases with increasing O/C ratio and it less sensitive to H/C ratio [35].

In order to evaluate exergetic efficiency of gasification of biofuels as function of fuel chemical composition, three different gasification temperature are considered: the carbon boundary temperature (when exactly enough oxygen is added to achieve complete gasification), and two fixed temperatures of 927°C and 1227°C which are beneficial to preserve the chemical exergy of the fuel in the product gas and required in practice to reduce kinetic limitations [36]. Gasification efficiency is evaluated for the adiabatic gasifier, which is fed by a biofuel, oxygen (both at the environmental conditions) and steam at 500 K and atmospheric pressure. The exergetic efficiency of the gasifier is evaluated as the ratio between exergy of the product gas diminished with exergy of oxygen and steam used as gasifying agent taken as the exergetic output and the exergy of converted biomass fuel taken as the exergetic input. Figure 5 shows the exergetic efficiencies of the gasifier for various O/C ratios in fuels [35].

The overall gasifier efficiency for all gasification temperatures decreases with increasing O/C ratio in a fuel. For fuels with low O/C ratio (e.g. coal $CH_{0.95}O_{0.2}$) the overall efficiency at fixed gasification temperatures (927 and 1227°C) is higher than that at the carbon boundary temperature. It is due to the steam effect in gasification, which is used to moderate gasification temperature, but the differences in efficiency are rather small. On the other hand for fuels with high O/C ratio (e.g. biomass $CH_{1.4}O_{0.6}$) the overall efficiency at the carbon boundary temperature is higher than that at fixed gasification temperatures. However, carbon boundary temperatures for biomass are too low from the point of view of reaction kinetics and biomass should be better gasified at higher temperatures.

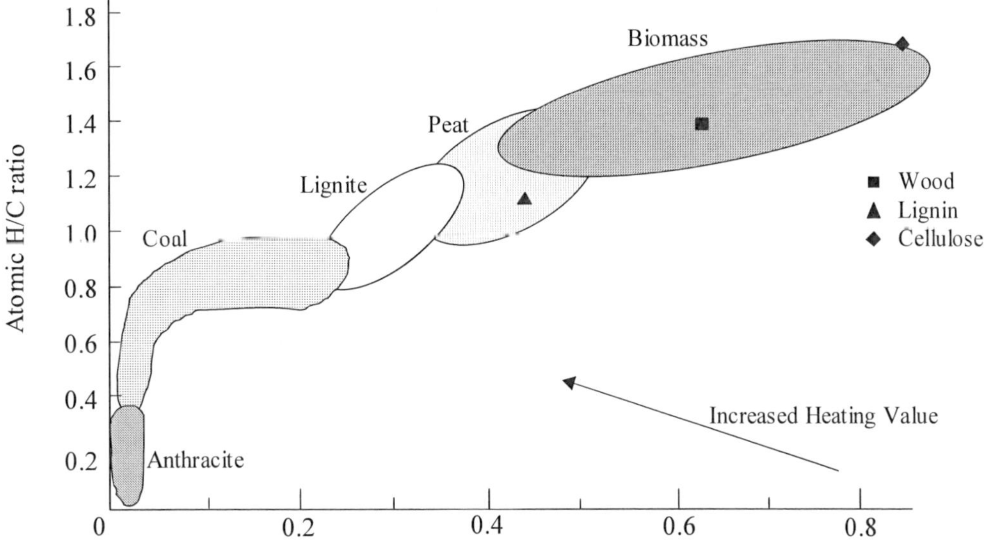

Figure 4. Van Krevelen diagram for solid biofuels.

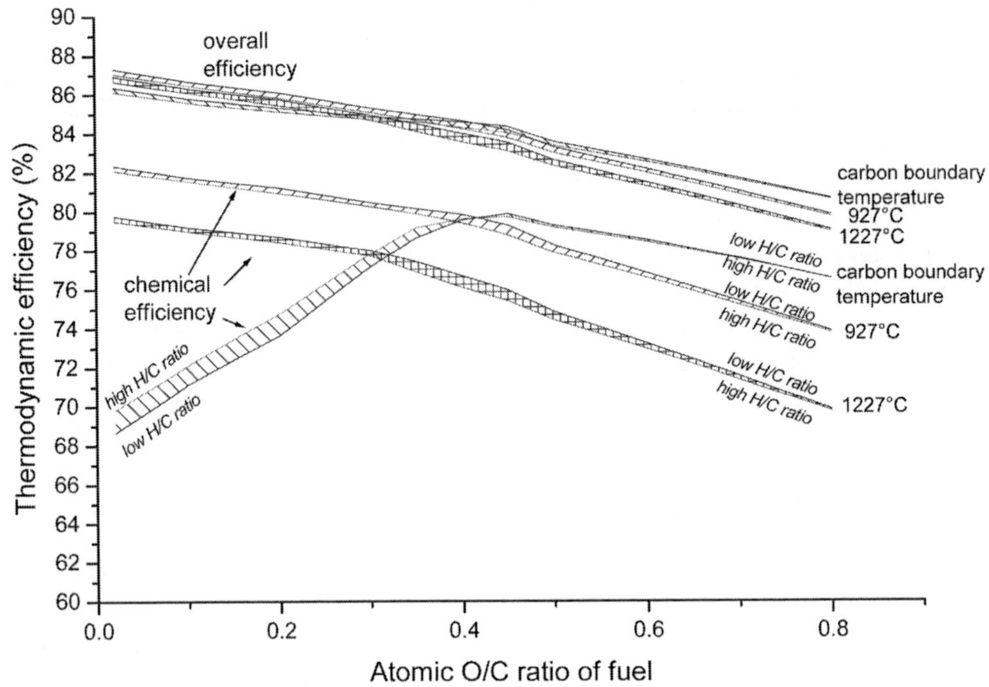

Figure 5. Exergetic efficiency of the gasifier for various biofuels.

The chemical efficiency, which involves only chemical exergy of biomass, increases with decreasing O/C ratio in a fuel only for gasification at fixed temperatures, whereas it shows maximum for gasification at the carbon boundary temperature. This maximum is due to a low contribution of chemical gas exergy at low O/C ratios (high carbon boundary temperature) and decrease of overall efficiency for high O/C ratios. For gasification at fixed temperatures 927° and 1227°C, the chemical efficiency in the O/C region below 0.4 is improved substantially due to moderation of the temperature with steam. Summarizing it can be concluded that the efficiency of biomass gasification can be improved by decreasing the O/C ratio in the fuel.

Biomass Torrefaction

Higher gasification efficiencies can be achieved for fuels with low O/C ratios, such as coal, than for fuels with high O/C ratios, such as wood, as explained in previous section. Therefore it is attractive to modify biomass properties by decreasing the ratio O/C and subsequently use the modified fuel as a gasifier feedstock. The O/C ratio in biomass can be lowered in a thermal pre-treatment process known as torrefaction where wood is treated in an inert gas atmosphere at temperature range 230-300°C [37-39]. Torrefied wood has a brown color and its properties are between wood and coal. Due to decomposition of hemicellulose the fibrous structure of wood is destructed and the hydrophilic nature of wood is inverted. This way grindability and fluidization properties of biomass are improved what is promising to implement large scale entrained-flow gasification of biomass which technology is currently used for coal.

Table 4. Composition of wood and torrefied wood (wt%)

	Wood	Torrefied wood (250°C, 30 min)	Torrefied wood (300°C, 10 min)
Carbon	47.2	51.3	55.8
Hydrogen	6.1	5.9	5.6
Oxygen	45.1	40.9	36.2
Nitrogen	0.3	0.4	0.5
Ash	1.3	1.5	1.9
LHV (MJ/kg)	17.6	19.4	21.0

During torrefaction the biomass partly devolatilises what leads to a decrease of mass, but initial energy content of the biomass is almost completely preserved in the solid product (fuel). The properties of torrefied wood depend on the type of wood used, the reaction temperature, and reaction time, usually 10-30 minutes. Typically, torrefied wood retains 70-90% of the original wood mass whereas the heating value increases by 5-25%. Table 4 summarizes the composition of wood and torrefied wood obtained for torrefaction of willow at different process temperatures and residence time [40]. As a result of torrefaction process, the O/C ratio of the wood decreases from 0.72 to 0.60 and 0.49, respectively, whereas the lower heating value increases from 17.6 to 19.4 and 21.0 MJ/kg, respectively.

Biomass torrefaction can be used as the pretreatment step before biomass gasification and different systems are possible. In the simplest system wood torrefaction is coupled with the circulating fluidized bed (CFB) gasifier, which is currently used for biomass gasification. Wood, which is torrefied at temperatures of 250 or 300°C, respectively, is used as a fuel in the gasifier. The CFB gasifier operates at 950°C and air is used as gasifying medium. In this system the volatiles from the torrefaction reactor are not utilized and regarded as a waste stream which should be cleaned. The hot product gases from the gasifier are physically quenched with cold gas to a temperature of 800°C. Steam produced in Heat Recovery Steam Generator (HRSG) is partially used to provide heat for the torrefaction process and partially exported. The pressure of the steam generated in the HRSG is tailored for controlling the torrefaction temperature: 45 bar for torrefaction at 250°C and 90 bar for torrefaction at 300°C.

In the more advanced system wood torrefaction can be coupled with entrained flow (EF) gasifier which is till now used only for coal gasification. Wood is torrefied at the same temperature range as in the previous system. Prior to its introduction into the gasifier, torrefied wood is pulverized to a particle size of 100 μm. The EF gasifier operates at 1200°C and oxygen is used as gasifying medium. In this system the volatiles from the torrefaction reactor are introduced in the top of the gasifier to induce a "chemical quench". The hot product gases from this gasifier are also physically quenched with cold gas to a temperature of 800°C to produce steam for the torrefaction reactor as well as export steam.

Exergetic efficiency is evaluated for all above-discussed torrefaction aided gasification system. The conventional biomass gasification of wood (without torrefaction) in CFB gasifier is used as a reference for efficiency comparison. Figure 6 shows the overall exergetic efficiency for various process options: I- conventional wood gasification in a CFB gasifier (no torrefaction), II – wood torrefaction and gasification of torrefied wood in a CFB gasifier, III – wood torrefaction and gasification of torrefied wood in a EF gasifier, whereas a and b refers to the torrefaction at temperature of 250 and 300°C, respectively.

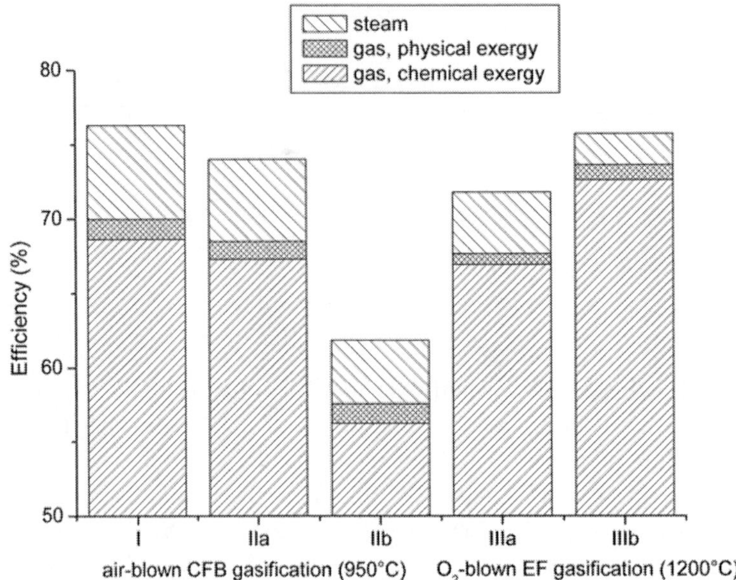

Figure 6. Exergetic efficiency for various torrefaction – gasification schemes.

In this figure the contributions of chemical and physical exergy of product gas as well as steam exergy to the overall efficiency are indicated. The overall efficiency of air-blown CFB gasification is lower for torrefied wood (cases IIab) than for wood (case I), especially at a high torrefaction temperature (case IIb) when a lot of energy is contained in the volatiles, which are not used in the process. The efficiency for oxygen-blown EF gasification of torrefied wood (case IIIb, torrefaction temperature $300^{o}C$) is quite comparable to that of air-blown CFB gasification of wood (case I), and slightly lower if torrefied wood is produced at $250°C$ (case IIIa). However, the highest amount of chemical exergy retained in the product gas is observed for case IIIb, that is torrefaction of $300^{o}C$ integrated with EF gasification at $1200^{o}C$. Summarizing, torrefaction is a novel promising method of biomass pre treatment provided heat produced in the gasifier is used for torrefaction reactions and volatiles from this process are utilized in the gasifier.

CONCLUSION

Biomass has great potential as renewable feedstock to produce biofuels. However, limitation of land and competition with food production are ecological drawbacks for large scale biomass applications. Therefore the most efficient technologies biomass-to-biofuels should be developed.

The most promising biomass-to-biofuels route is biomass gasification coupled with synthesis of biofuels, including Fischer-Tropsch hydrocarbons, hydrogen, and methanol. Exergy analysis shows that the largest process losses in the overall biomass-to-biofuel chains are due to biomass gasification.

The efficiency of biomass gasification can be improved by operating gasifier at the optimal conditions that is at the carbon boundary point, where enough exact oxygen is added to complete gasification. Moreover, reduction of O/C ratio in biomass can improve the exergetic efficiency. Torrefaction is a novel promising method of biomass pre treatment provided heat produced in the gasifier is used for torrefaction reactions and volatiles from this process are utilized in the gasifier.

REFERENCES

[1] Birol F, Argiri M. World energy prospects to 2020. *Energy* 1999, *24*, 905-918.

[2] Smalberg F, Moulijn J, van Bekkum H. Global carbon dioxide emission and consumption. *Green Chemistry* 2000, *2*, G97-G100.

[3] Easterly JL, Burnham M. Overview of biomass and waste fuel resources for power production. *Biomass Bioenergy* 1996, *10*, 79-92.

[4] Chum HL, Overend RP. Biomass and renewable fuels. *Fuel Processing Technology* 2001, *71*, 187-195.

[5] Huber GW, Iborra S, Corma A. Synthesis of transportation fuels from biomass: chemistry, catalysts, and engineering. *Chem. Rev.* 2006, *106*, 4044-4098.

[6] Hanegraaf MC, Biewinga EE, van der Bijl G. Assessing the ecological and economic sustainability of energy crops. *Biomass Bioenergy* 1998, *15*, 345-355.

[7] Herman WA. Quantifying global exergy resources. *Energy* 2006, *31*, 1349-1366.

[8] Szargut J, Morris DR, Steward FR. *Exergy analysis of thermal, chemical, and metallurgical processes*. New York: Hemisphere Publishing Corporation, 1988.

[9] Dincer I, Rosen MA. Thermodynamic aspects of renewable and sustainable development. *Renewable Sustainable Energy Rev.* 2005, *9*, 169-189.

[10] Devi L, Nair SA, Pemen AJM, Yan K, van Heesch EJM, Ptasinski KJ, Janssen FJJG. Tar removal from biomass gasification processes, In: Columbus F, Ed. *Biomass and Bioenergy: New Research*. Hauppauge, NY: NOVA Science Publishers, 2006.

[11] Warnecke R. Gasification of biomass: comparison of fixed bed and fluidized bed gasifier. *Biomass Bioenergy* 2000, *18*, 489-497.

[12] Sutton D, Kelleher, Ross JRH. Review of literature on catalysts for biomass gasification. *Fuel Processing Technology* 2001, *73*, 155-173.

[13] Ptasinski KJ. Exergy analysis of a combined fuel processor and fuel cell sysytem. In: Alemo P, Ed. *Progress in Fuel Cell Research*. Hauppauge, NY: Nova Science, 2007.

[14] Kotas TJ, *The exergy method for thermal plant analysis*, Krieger: Malabar, 1995.

[15] Huber GW, Dumesic JA. An overview of aqueous-phase catalytic processes for production of hydrogen and alkanes in a biorefinery. *Catalysis Today*, 2006, *111*, 119-132.

[16] Lapidus A, Krylova A, Paushkin Y, Rathousky J, Zukal A, Starek J. Synthesis of liquid fuels from products of biomass gasification. *Fuel* 1994, *73*, 583-590.

[17] Ptasinski KJ. Efficiency analysis for the production of modern energy carriers from renewable resources and wastes. In: Tiezzi E, Marquez JC, Brebbia CA, Jørgensen SE, Eds. *Ecosystems and Sustainable Development VI*. Southampton: WIT Press, 2007, p. 239-249.

[18] Demirbas A. Biomass resource facilities and biomass conversion processing for fuels and chemicals. *Energy Conversion Management* 2001, *42*, 1357-1378.

[19] Boerrigter H, den Uil H. Green diesel from biomass with the Fischer-Tropsch synthesis. In: Palz W, Spitzer J, Maniatis K, Kwant K, Helm P, Grassi A, Eds. *Proceedings 12th European Biomass Conference, Amsterdam, The Netherlands.* Florence: ETA-Florence and WIP-Munich, 2002 p, 1152-1153.

[20] Schulz H. Short history and present trends of Fischer–Tropsch synthesis. *Applied Catalysis A: General* 1999, *186*, 3-12.

[21] Prins MJ, Ptasinski KJ, Janssen FJJG. Exergetic optimization of a production process of Fischer-Tropsch fuels from biomass. *Fuel Processing Technology* 2004, *86*, 375-389.

[22] Espinoza RL, Steynberg AP, Jager B, Vosloo AC. Low temperature Fischer-Tropsch synthesis from a Sasol perspective. *Applied Catalysis A: General* 1999, *186*, 13-26.

[23] Conte M, Iacobazzi A, Ronchetti M, Vellone R. Hydrogen economy for a sustainable development: state-of-the-art and technological perspectives. *J. Power Sources* 2001, *100*, 171-187.

[24] Veziroglu TN, Momirlan M. Current status of hydrogen energy. *Renewable Sustainable Energy Rev.* 2001, *6*, 141-179.

[25] Dincer I. Technical, environmental and exergetic aspects of hydrogen energy systems. *Int. J. Hydrogen Energy* 2002, *27*, 265-285.

[26] Ptasinski KJ, Prins MJ, van der Heijden S. Exergy analysis of hydrogen production methods from biomass In: Frangopoulos CA, Rakopoulos, Tsatsaronis G, Eds. *Proceedings 19th International Conference on Efficiency, Cost, Optimization, Simulation and Environmental Impact of Energy Systems ECOS 2006, Aghia Pelagia, Greece.* Athens: NTUA, 2006, p. 1601-1608.

[27] Anikeev VI, Gudkov AV, Ermakova A. Exergetic analysis of biomass gasification for co-production of methanol and energy. *Theoretical Foundations Chemical Engineering* 1996, *30*, 461-468.

[28] Ptasinski KJ, Hamelinck C, Kerkhoff PJAM. Exergy analysis of methanol from sewage sludge process, *Energy Conversion Management* 2002, *43*, 1445-1457.

[29] Rosen MA, Thermodynamic investigation and comparison of selected production processes for hydrogen and hydrogen-derived fuels. *Energy* 1996, *21*, 1079-1094.

[30] Rosen MA, Scott DS. Comparative efficiency assessments for a range of hydrogen production processes. *Int. J. Hydrogen Energy* 1998, *23*, 653-659.

[31] Cairns EJ, Tevebaugh AD. CHO gas phase compositions in equilibrium with carbon, and carbon deposition boundaries at one atmosphere. *J. Chem. Eng. Data* 1964, *9*, 453-462.

[32] Prins MJ, Ptasinski KJ, Janssen FJJG. Thermodynamics of gas-char reactions: first and second law analysis. *Chem. Eng. Science* 2003, *58*, 1003-1011.

[33] Li X, Grace JR, Watkinson AP, Lim CJ, Ergüdenler CJ. Equilibrium modeling of gasification: a free energy minimization approach and its application to a circulating fluidized bed coal gasifier. *Fuel* 2001, *80*, 195-207.

[34] van Krevelen DW. *Coal: typology, physics, chemistry, constitution.* Amsterdam: Elsevier Science Publishers B.V., 1993.

[35] Prins MJ, Ptasinski KJ, Janssen FJJG. From coal to biomass gasification: comparison of thermodynamic efficiency. *Energy* 2007, *32*, 1248-1259.

[36] Desrosiers R. Thermodynamics of gas-char reactions. In: Reed TB, Ed. *A Survey of Biomass Gasification*. Colorado: Solar Energy Research Institute, 1979.

[37] Bourgois J, Guyonnet R. Characterization and analysis of torrefied wood. *Wood Sci. Technology* 1988, *22*, 143-155.

[38] Prins MJ, Ptasinski KJ, Janssen FJJG. Torrefaction of wood. Part I: Weight loss kinetics. *J. Analytical Applied Pyrolysis* 2006, *77*, 28-34.

[39] Prins MJ, Ptasinski KJ, Janssen FJJG. Torrefaction of wood. Part II: Analysis of products. *J. Analytical Applied Pyrolysis* 2006, *77*, 35-40.

[40] Prins MJ, Ptasinski KJ, Janssen FJJG. More efficient biomass gasification via torrefaction. *Energy* 2006, *31*, 3458-3470.

In: New Research on Biofuels

ISBN 978-1-60456-828-8

Editors: J. H. Wright and D. A. Evans

© 2008 Nova Science Publishers, Inc.

Chapter 6

THE ENERGY BALANCE AND FUEL PROPERTIES OF BIODIESEL

Mustafa Acaroğul[*] and Mahmut Ünaldı

Selçuk University, Technical Education Faculty Kampüs, Konya, Turkey

ABSTRACT

In this study energy balance and fuel properties of biodiesel has been calculated. Accordingly, the cost of 1 liter of oil is calculated 0.32 € after the income from the seed meal is deduced. Finally, the cost of per unit of biodiesel (1 liter) was calculated as 0.55 €, after deduction of the income provided by the sales of glycerin for use in soap and cosmetic industry.

The energy equivalent of total output was calculated 147605.50 MJ per hectare. The net energy gain (refined oil) was found as 15105.63 MJ per hectare (The net energy ratio 11.031) according to yield and inputs values.

The viscosity values of vegetable oils vary between 27.2 and 53.6 mm^2/s whereas those of vegetable oil methyl esters between 3.59 and 4.63 mm^2/s. The flash point values of vegetable oil methyl esters are highly lower than those of vegetable oils. The flash point values of vegetable oil methyl esters are highly lower than those of vegetable oils. An increase in density from 860 to 885 kg/m^3 for vegetable oil methyl esters or biodiesel increases the viscosity from 3.59 to 4.63 mm^2/s and the increases are highly regular. There is high regression between density and viscosity values vegetable oil methyl esters. The relationships between viscosity and flash point for vegetable oil methyl esters are irregular. An increase in density from 860 to 885 kg/m^3 for vegetable oil methyl esters increases the flash point from 401 to 453 K and the increases are slightly regular.

The LHV values of vegetable oils methyl ester vary between 35.74 and 39.16 MJ/kg.

Keywords: biodiesel, fuel, energy balance

---

[*] Corresponding author: Assoc. Prof. Dr. acaroglu@selcuk.edu.tr Tel: 90-332-2233321 Fax:90-332-2412179

1. INTRODUCTION

Biomass is a renewable resource so far as its production is continued in a sustainable way. Biomass conversion is a process to convert photosynthetic material into a more useful form. Fuels from biomass can take various forms such as solid, gas, and liquid. Liquid biomass (Vegetable oils) from rapeseed, safflower, soybean, palm oil, sunflower, and others can be used for diesel engine. Vegetable oils are used for food, as industrial raw materials, and for the generation of energy. Vegetable oils are a blend of free fatty acids (FFA's); monoglycerides, diglycerides, and triglycerides; phosphatides, lipoproteins, and glycolipids; waxes; terpenes, gums, and other less important compounds. Today's diesel engines require a clean-burning, stable fuel that performs well under a variety of operating conditions. Biodiesel is the only alternative fuel that can be used directly in any existing, unmodified diesel engine. Because it has similar properties to petroleum diesel fuel, biodiesel can be blended in any ratio with petroleum diesel fuel [1-3, 13].

Biodiesel is the name for a variety of ester-based oxygenated fuels made from vegetable oils or animal fats. Biodiesel is the name for a variety of ester-based oxygenated fuels made from vegetable oils or animal fats. The concept of using vegetable oil as a fuel dates back to 1895 when Dr. Rudolf Diesel developed the first diesel engine to run on vegetable oil. Biodiesel can be produced by several processes. Micro emulsification, pyrolysis transesterification and super critical method are the four different techniques used to production biodiesel. Vegetable oils or fats can be converted to fatty acids, which in turn are converted to esters. Oils or fats can also be converted to methyl or ethyl esters directly, using an acid or base to accelerate (catalyze) the transesterification reaction. Biodiesel is produced through a process known as transesterification, as shown in the equation below, where R1, R2, and R3 are long hydrocarbon chains, sometimes called fatty acid chains. There are only five chains that are most common in soybean oil and animal fats (others are present in small amounts) (Figure I) [4, 6, 8-11, 16-30].

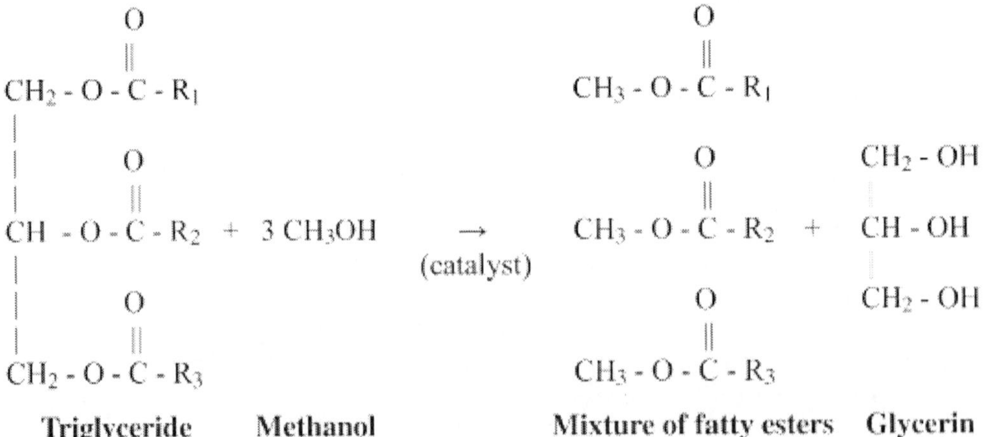

Figure 1. Transesterification flow chart [1, 16].

Base catalyzation is preferred, because the reaction is quick and thorough. It also occurs at lower temperature and pressure than other processes, resulting in lower capital and operating costs for the biodiesel plant. The most common method of producing biodiesel is to reaction animal fat or vegetable oil with methanol in the presence of sodium hydroxide (a base, known as lye or caustic soda). This reaction is a base-catalyzed transesterification that produces methyl esters and glycerin. Methanol is preferred, because it is less expensive than ethanol. The majority of the alkyl esters produced today is done with the base catalyzed reaction because it is the most economic for several reasons:

- Low temperature (65 ^{0}C) and pressure (1.37*10^5 Pascal) processing.
- High conversion (98%) with minimal side reactions and reaction time.
- Direct conversion to methyl ester with no intermediate steps.

The use of biodiesel in a conventional diesel engine results in substantial reduction of unburned hydrocarbons, carbon monoxide, and particulate matter. Emissions of nitrogen oxides are either slightly increased depending on the duty cycle and testing methods. The use of biodiesel decreases the solid carbon fraction of particulate matter (since the oxygen in biodiesel enables more complete combustion to CO_2), eliminates the sulphate fraction (as there is no sulphur in the fuel), while the soluble, or hydrocarbon, fraction stays the same or is increased. Therefore, biodiesel works well with new technologies such as catalysts (which reduces the soluble fraction of diesel particulate but not the solid carbon fraction), particulate traps, and exhaust gas recirculation (potentially longer engine life due to less carbon). Biodiesel is less toxic than petroleum diesel and biodegrades as fast as dextrose. In addition, biodiesel has a flash point of over 125°C which makes it safer to store and handle than petroleum diesel fuel [15-28, 30, 31].

2. FUEL PROPERTIES OF BIODIESEL

Vegetable oils and their derivatives (especially methyl esters), commonly referred to as "biodiesel," are prominent candidates as alternative diesel fuels. Biodiesel is generally made of methyl esters of fatty acids produced by the transesterification reaction of triglycerides with methanol in the presence alkali as a catalyst [8]. Among the alcohols that can be used in the transesterification reaction are methanol, ethanol, propanol, butanol and amyl alcohol. Methanol and ethanol are used most frequently, ethanol is a preferred alcohol in the transesterification process compared to methanol because it is derived from agricultural products and is renewable and biologically less objectionable in the environment, however methanol because of its low cost and its physical and chemical advantages. The transesterification reaction can be catalyzed by alkalis [9, 10, 40, 61], acids [15], or enzymes [11, 18, 44, 46, 54].

The transesterification of triglycerides by supercritical methanol, ethanol, propanol and butanol, has proved to be the most promising process [1, 36, 37]. Supercritical methanol has a high potential for both transesterification of triglycerides and methyl esterification of free fatty acids to methyl esters for diesel fuel substitute. In the supercritical methanol transesterification method, the yield of conversion raises 95% for 10 minutes.

The catalyst (NaOH) is dissolved into methanol by vigorous stirring in a small reactor. The oil is transferred into the biodiesel reactor and then the catalyst/alcohol mixture is pumped into the oil. The final mixture is stirred vigorously for 1 hour at 338 K in ambient pressure. A successful transesterification reaction produces two liquid phases: ester and crude glycerin. Crude glycerin, the heavier liquid, will collect at the bottom after several hours of settling. Phase separation can be observed within 10 minutes and can be complete within 2 hours of settling. Complete settling can take as long as 20 hours. After settling is complete, water is added at the rate of 5.5 percent by volume of the methyl ester of oil and then stirred for 5 minutes and the glycerin is allowed to settle again. Washing the ester is a two-step process, which is carried out with extreme care. A water wash solution at the rate of 28 percent by volume of oil and 1 gram of tannic acid per liter of water is added to the ester and gently agitated. Air is carefully introduced into the aqueous layer while simultaneously stirring very gently. This process is continued until the ester layer becomes clear. After settling, the aqueous solution is drained and water alone is added at 28 percent by volume of oil for the final washing [1, 35, 36, 39, 41-43, 52-54, 57-61].

All the runs of supercritical methanol transesterification were performed in a 100-mL cylindrical. The sample was loaded from the bolt-hole into the autoclave, and the hole was plugged with a screw bolt after each run. In a typical run, the autoclave was charged with a given amount of vegetable oil (20-30 g) and liquid methanol (30-50 g) with changed molar ratios. The autoclave was supplied with heat from an external heater, and power was adjusted to give an approximate heating time of 15 min. The temperature of the reaction vessel was measured with an iron-constantan thermocouple and controlled at ±5 K for 30 min. Transesterification occurred during the heating period.

Eight different samples of biodiesel were used for viscosity, flash point and density measurements. A Redwood No. 1 viscosity meter with a measuring cup and a thermostat was used to measure the viscosity of all samples. The viscosity measurements were carried out at 313 K temperature. The temperatures were checked with a digital thermometer within the thermostat and the viscosity meter. At the beginning of each measurement a volume of 50 ml of the sample was filled into the measuring cup. We had to adjust shear rates for the different kinds of samples, because the viscosities are quite different and the viscosity meter has to be used within the correct measuring range. Flash point measurements were carried out using a Koehler mark apparatus.

Table 1. Fatty acid compositions of vegetable oil samples [1, 10]

Sample	16:0	16:1	18:0	18:1	18:2	18:3	Others
Cottonseed	28.7	0	0.9	13.0	57.4	0	0
Rapeseed	3.5	0	0.9	64.1	22.3	8.2	0
Safflowerseed	7.3	0	1.9	13.6	77.2	0	0
Safflowerseed	6.4	0.1	2.9	17.7	72.9	0	0
Palm	42.6	0.3	4.4	40.5	10.1	0.2	1.1
Soybean	13.9	0.3	2.1	23.2	56.2	4.3	0
Hazelnut kernel	4.9	0.2	2.6	83.6	8.5	0.2	0

Table 2. Viscosity, density and flash point measurements of ten vegetable oils [1, 17]

Oil Source	Viscosity mm^2/s (at 311 K)	Density kg/m^3	Flash Point K
Corn	34.9	909.5	550
Cottonseed	33.5	914.8	509
Crambe	53.6	904.4	447
Linseed	27.2	923.6	514
Peanut	39.6	902.6	544
Rapeseed	37.0	911.5	519
Safflower	31.3	914.4	533
Sesame	35.5	913.3	533
Soybean	32.6	913.8	527
Sunflower	33.9	916.1	447

Table 3. Some fuel properties of six methyl ester biodiesels

Source	Viscosity cSt at 313.2 K	Density g/mL at 288.7 K	Cetane Number	Reference
Sunflower	4.6	0.880	49	[49]
Soybean	4.1	0.884	46	[53]
Palm	5.7	0.880	62	[49]
Penaut	4.9	0.876	54	[57]
Babassu	3.6	--	63	[57]
Tallow	4.1	0.877	58	[4]

Table 4. Viscosity, density and flash point measurements of eight oil methyl esters [3]

Methyl ester	Viscosity mm^2/s (at 313 K)	Density kg/m^3 (at 288 K)	Flash point K
Cottonseed oil	3.69	880	437
Hazelnut kernel oil	3.59	860	401
Mustard oil	4.10	881	446
Palm oil	3.70	870	443
Rapeseed oil	4.63	885	428
Safflower oil	4.03	880	453
Soybean oil	4.08	885	447
Sunflower oil	4.22	880	443

Table 5. LHV (Low Heating Value) properties of methyl ester biodiesel[*]

Methyl ester	LHV (MJ/kg)	Ash (wt %)	Moisture (wt %)
Cottonseed oil	39.16		
Hazelnut kernel oil	38.70	0.013	0.75
Mustard oil	39.04	--	--
Palm oil	35.74	0.003	6.30
Rapeseed oil	38.24	0.035	1.30
Safflower oil	36.98	0.004	0.61
Sunflower oil	35.92	0.002	0.90

* Analysis time July 2005 (Acaroğlu).

3. BIODIESEL EMISSIONS COMPARED TO CONVENTIONAL DIESEL

Biodiesel is the first and only alternative fuel to have a complete evaluation of emission results and potential health effects submitted to the U.S. Environmental Protection Agency (EPA) under the Clean Air Act Section 211 (b). These programs include the most stringent emissions testing protocols ever required by EPA for certification of fuels or fuel additives in the US. The data gathered through these tests complete the most thorough inventory of the environmental and human health effects attributes that current technology will allow.

The overall ozone (smog) forming potential of biodiesel is less than diesel fuel. The ozone forming potential of the speciated hydrocarbon emissions was nearly 50 percent less than that measured for diesel fuel.

Sulfur emissions are essentially eliminated with pure biodiesel. The exhaust emissions of sulfur oxides and sulfates (major components of acid rain) from biodiesel were essentially eliminated compared to sulfur oxides and sulfates from diesel.

Criteria pollutants are reduced with biodiesel use. The use of biodiesel in an unmodified Cummins N14 diesel engine resulted in substantial reductions of unburned hydrocarbons, carbon monoxide, and particulate matter. Emissions of nitrogen oxides were slightly increased.

Carbon Monoxide - The exhaust emissions of carbon monoxide (a poisonous gas) from biodiesel were 47 percent lower than carbon monoxide emissions from diesel.

CO_2 emission index is defined as the CO_2 emission (%) divided by corresponding fuel consumption rate (in unit of g/h), the CO emission index is defined as CO emission (ppm) divided by the corresponding fuel consumption rate (in unit of g/h) and the NOx emission index is defined as NOx emission (ppm) divided by the corresponding fuel consumption rate. The catalytic converter reduced CO and HC emissions [1].

Benzene emissions increased with the amount of RME (rapeseed oil methylester) either with catalytic converter or without catalytic converter. This is remarkable because benzene is absent in RME. This indicates that the main source of benzene emissions may be a synthesis that occurs during combustion, rather than unburned fuel. However, the catalytic converter reduced the emissions by one third.

Particulate Matter - Breathing particulate has been shown to be a human health hazard. The exhaust emissions of particulate matter from biodiesel were 47 percent lower than overall particulate matter emissions from diesel,

Hydrocarbons - The exhaust emissions of total hydrocarbons (a contributing factor in the localized formation of smog and ozone) were 67 percent lower for biodiesel than diesel fuel.

Nitrogen Oxides(NOx) emissions from biodiesel increase or decrease depending on the engine family and testing procedures. NOx emissions (a contributing factor in the localized formation of smog and ozone) from pure (100%) biodiesel increased in this test by 10 percent. However, biodiesel's lack of sulfur allows the use of NOx control technologies that cannot be used with conventional diesel. So, biodiesel NOx emissions can be effectively managed and efficiently eliminated as a concern of the fuel's use [14, 26, 31, 33, 38, 49, 50, 59].

Biodiesel reduces the health risks associated with petroleum diesel. Biodiesel emissions showed decreased levels of PAH and nitrited PAH compounds which have been identified as potential cancer causing compounds. In the recent testing, PAH compounds were reduced by 75 to 85 percent, with the exception of benzo(a)anthracene, which was reduced by roughly 50 percent. Targeted nPAH compounds were also reduced dramatically with biodiesel fuel, with 2-nitrofluorene and 1 -nitropyrene reduced by 90 percent, and the rest of the nPAH compounds reduced to only trace levels.

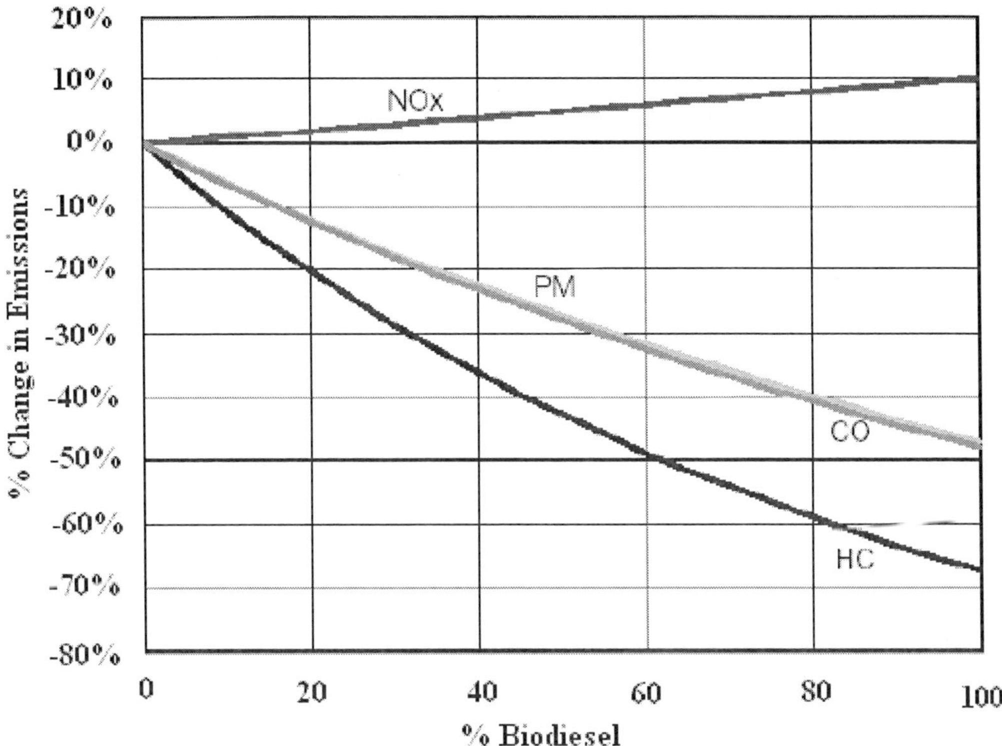

Figure 2. Average emission impacts of biodiesel fuels in CI engines [59].

4. THE ENERGY BALANCE OF SAFFLOWER OIL

Energy analysis, along with economic and environmental analyses, is an important tool to define the behavior of agricultural system. Energy analysis started as a relevant subject in agricultural production in the 1970's as a results of the dramatic increase of oil derivative prices. In an energy analysis of production systems it is necessary to consider the following steps ([1, 32, 47]:

- Set a limit in the process or system to be analyzed in such a way that all inputs and outputs, which pass that limit in a certain time interval, are evaluated.
- Assign energy requirements to all inputs
- Identify and quantify all outputs, establishing criteria for energy embodied in the main products and that corresponding to by-products.
- Relate output energy to total sequestered energy to obtain the energy ratio and the energy productivity.
- Apply energy analysis results

Safflower, Carthamus tinctorius, is among the oldest crops known to man. The *Safflower* (*Carthamus tinctorius L.*) is probably native to an area bounded by the Eastern Mediterranean and Persian Gulf. Believed to have originated in southern Asia and is known to have been cultivated in China, India, Persia and Egypt almost from prehistoric times. During middle Ages it was cultivated in Italy, France, and Spain, and soon after discovery of America, the Spanish took it to Mexico and then to Venezuela and Colombia. It was introduced into United States in 1925 from the Mediterranean region and is now grown in all parts west of 100th meridian. *Safflower* is commercially cultivated to a large extent in India, Mexico and the USA (Table 6, Table 7) [5, 12, 34, 48].

Safflower is cultivated for the edible oil obtained from the seed. It contains a higher percentage of essential unsaturated fatty acids and a lower percentage of saturated fatty acids than other edible vegetable seed oils. *Safflower* oil lowers blood cholesterol levels and is used to treat heart diseases. The flowers have been the source of yellow and red dyes, largely replaced by synthetics, but still used in rouge. Seeds used for tumors, especially inflammatory tumors of the liver [19].

Table 6. Safflower in world (country, sown area, production and yield) [29]

Country	Sown Area (ha)	Production (ton)	Yield (kg/ha)
Indian	404100	226000	559
USA	79320	135160	1704
Ethiopia	72000	38000	528
Mexico	52758	52855	1002
Australia	35000	26000	743
Turkey	30	15	500
World	756055	577555	764

Table 7. Safflower in Turkey (year, sown area, production and yield) (1995-2002) [29]

Years	Sown area (ha)	Production (ton)	Yield (kg/ha)
1995	134	125	930
1996	81	74	910
1997	74	65	880
1998	75	72	960
1999	50	50	1000
2000	30	18	600
2001	35	25	714
2002	30	15	500
Konya (2003-2004)	35	67.375	1925

Flowers considered diaphoretic, emmenagogue, laxative, sedative, stimulant, in large doses laxative; used as a substitute or adulterant for saffron in treating measles, scarlatina, and other exanthematous diseases. Charred *safflower* oil used for rheumatism and sores; seeds, diuretic and tonic [7]. In China, prescribed as uterine astringent in dysmenorrhea. In Iran, the oil is used as a salve for sprains and rheumatism.

Safflower is propagated by seed. The seed is sown 2-3 cm deep, with planting distances 10 cm in the row and 30-60 cm between the rows. The desired crop density is 40-50 plants per m^2, and up to 70 plants per m^2 on light soil. The growing period is 200-250 days or 110-140 days, respectively [5, 48, 51].

Table 8. The properties of *safflower* variety crops [5]

Sort (variety)	Thorny	Flower Color	Plant height (cm)	Seed color	Oil Content (%)	Weight (gr/1000 seed)
Yenice	Not Thorny	Red	100-120	White	24-25	38-40
Dinçer	Not thorny	Orange	90-110	White	25-28	45-49
5-154	Thorny	Yellow	60-80	White	35-40	46-50

Table 9. The properties on sowing operations

Operation	Properties
Row distance	15 cm
Row upper	10 cm
Sowing deep	2.5-4.0 cm
Sowing (seeding) norm	20 kg/ha
Crop density	$66 - 67$ plant per m^2
Fertilizer (N)	120 kg/ha per year
Fertilizer (P_2O_5)	50 kg/ha per year
Fertilizer (K)	Not used

No recommendations can be made for herbicide use, as this aspect of weed control is still being tested, but herbicides appropriate for sunflower crops are believed to be appropriate for *safflower* too [12]. The yields of seed can reach 1.1 – 1.7 t/ha/year [29]. With irrigation and good fertilization, yields 2.8-4.5 t/ha/year can be achieved, but the world average yield is 0.5 t/ha/year. The cultivars have 30-48 % oil in the fruit. The oil contains 73-79 % linoleic acid, depending on the cultivar [29, 45, 55, 56, 60].

In Turkey, today there are only three Safflower kinds. This study was used Dinçer variety (Table 8). Middle Anatolia region was selected and 35 ha Safflower was sowed in Konya. Sowing time was the first week of April (Table 9).

Harvesting is making with Combine harvester.

5. RESULTS AND DISCUSSION

In this study energy balance in Safflower production has been determined and calculated (Table 10 and Table 11). In making the energy balance, input and output values used in calculation are measured in field conditions and calculated using literature. Table 5 list the cultivation operations carried out, with the corresponding work hours and technical means and used with their respective consumption.

Table 10. Technical characteristic of machines and energy inputs *safflower* (MJ/ha)

Agricultural Operation	(h/ha)	Tractors Energy	Equipment Energy	Fuel- lubricant Energy	Labor Energy	TOTAL
Primary tillage plough	2.14	73.36	14.35	604.53	4.00	696.24
Secondary tillage (disk harrow)	0.84	28.91	10.16	238.33	1.58	278.97
Seeding	0.68	23.34	49.94	192.47	1.27	267.03
Fertilizer	0.22	7.55	1.96	62.18	0.41	72.11
Harvesting (combine)	0.94		121.20	390.74	1.76	513.70
Total		133.16	197.62	1488.25	9.01	1828.04

Table 11. Fertilizer and seed energy inputs

Fertilizer Energy	kg/ha	MJ/kg	MJ/ha
N	120	49.1	5892
P	40	17.78	711.2
		TOTAL	6603.2
Seed Energy	20	14	280

Table 12. Total energy inputs of *safflower*

Energy Input	MJ/ha	%
Fuel-oil	1488.25	12.13
Tractor Energy	133.16	1.09
Machinery Energy	197.62	1.61
Labour Energy	9.01	0.07
Fertilizer Energy	6603.20	53.83
Seed Energy	280.00	2.28
Industrial process (extraction)	2406.25	19,61
Industrial process (refining)	1150.38	9.38
TOTAL	12267.87	100.00

Table 13. Total energy outputs of *safflower*

Energy Output	MJ/kg	MJ/ha
Safflower oil (Refined)	39.5	27373.5
Safflower stalk	18.08	120232
Safflower product residues	18.09	28871.6
Total		147605.5

Table 14. Input and cost *safflower* biodiesel

Raw material	Cost per liter (€)	Conversion %	Quantity / liter biodiesel	Cost / liter biodiesel (€)
Oil	0.32	90	1.11	0.36
Methanol	0.250	-	0.20	0.05
NaOH	1.200	-	0.035	0.04
Various	-	-	-	0.01
Energy	0.100	-	-	0.10
Biodiesel cost	-	-	-	0.55

In harvesting, the Safflower seed yield was obtained as 1925 kg per hectare on an average and *Safflower* stalk was obtained as 6650 kg DM per hectare in the moisture of 15 % (18.08 MJ/kg). The average seed oil content was found 36 % (693 kg per hectare oil yield), (39.50 MJ/kg). (Table 12, Table 13).

Cost inputs in safflower oil biodiesel production has been calculated (table 14). In making the cost analysis, input and output values used in calculation are measured in process conditions and calculated using literature.

The energy equivalent of total output was calculated 147605.50 MJ per hectare. The net energy gain (refined oil) was found as 15105.63 MJ/ha (The net energy ratio 11.031) according to yield and inputs values. Net energy gain of Safflower seed is higher then other oil seeds in the Turkey. In addition to these, Safflower crops will have a great importance in

Turkey as oil resource and biomass energy source as it solves environmental problems, its inputs are low, and it can be grown in arid zones.

6. REFERENCES

[1] Acaroğlu, M. 2003. *Renewable Energy Sources,* Atlas Yayın & Dağıtım, Istanbul, (Turkish Book)

[2] Acaroğlu, M., Gezer, I., 2004, Renewable Energy Sources and Future of Biomass Energy in Turkey, pp. 413-416, *2nd World Conference on Biomass for Energy, Industry and Climate Protection,* 10-14 May 2004, Rome, Italy

[3] Acaroğlu, M., Demirbaş, A., 2005, *Relationships between Viscosity and Density Measurements of Biodiesel Fuels, Energy Sources* (in press).

[4] Ali Y, Hanna MA, Cuppett SL. 1995. Fuel properties of tallow and soybean oil esters. JAOCS 1995; 72:1557-1564.

[5] Babaoğlu, M., 2004 *Dünya'da ve Türkiye'de Aspir Bitkisinin Tarihçesi, Kullanım Alanları ve Önemi* (Turkish), Trakya Tarımsal Araştırma Enstitüsü Edirne.

[6] Bala, B. K. 2005. Studies on biodiesels from transformation of vegetable oils for diesel engines, *Energy Edu. Sci. Technol.* 15:1-43.

[7] C.S.I.R. (Council of Scientific and Industrial Research), 1948-1976, *The wealth of India.* 11 vols, New Delhi.

[8] Clark,S. J., L. Wagner, M. D. Schrock, and P. G. Pinnaar. 1984. Methyl and ethyl esters as renewable fuels for diesel engines. *JAOCS* 61:1632–1638.

[9] Demirbas A. Biodiesel from vegetable oils via transesterification in supercritical methanol, *Energy Convers Mgmt* 2002; 43:2349–56.

[10] Demirbas A. Biodiesel fuels from vegetable oils via catalytic and non-catalytic supercritical alcohol transesterifications and other methods: a survey. *Energy Convers Management* 2003;44:2093-2109.

[11] Du W, Xu Y, Liu D, Zeng J. Comparative study on lipase-catalyzed transformation of soybean oil for biodiesel production with different acyl acceptors, *J Molecular Catal B: Enzymatic* 2004;30:125–129.

[12] Duke, J.A., 1983. *Handbook of Energy Crops* (unpublished).

[13] El Bassam, N., 1998. *Energy Plant Species, Their use and impact on environment, James&James* (Science Publishers), 320 p., ISBN 1 873936 75 3.

[14] EPA, 2002, *A Comprehensive Analysis of Biodiesel Impacts on Exhaust Emissions,* Draft Technical Report, EPA420-P-02-001 October 2002.

[15] Furuta S, Matsuhashi H, Arata K. Biodiesel fuel production with solid superacid catalysis in fixed bed reactor under atmospheric pressure. *Catal Commun* 2004; 5:721–723.

[16] Gerpen, V., 2005, Biodiesel processing and production, *Fuel Processing Technology 86* (2005) 1097–1107.

[17] Goering, E., W. Schwab, J. Daugherty, H. Pryde, and J. Heakin, 1982, Fuel properties of eleven vegetable oils. *Trans ASAE* 25: 1472–1483.

[18] Hama S, Yamaji H, Kaieda M, Oda M, Kondo A, Fukuda H. Effect of fatty acid membrane composition on whole-cell biocatalysts for biodiesel-fuel production. *Biochem Eng J* 2004;21:155–60

[19] Hartwell, J.L. 1967-1971. Plants used against cancer. *A survey. Lloydia* 30-34.

[20] http://journeytoforever.org/biodiesel.html

[21] http://ww2.green-trust.org:8383/biodiesel2.htm

[22] http://www.biodiesel.org

[23] http://www.biodiesel.org/pdf_files/fuelfactsheets/emissions.pdf

[24] http://www.canentec.com/whatisbiodiesel.html

[25] http://www.distributiondrive.com/Article16.html

[26] http://www.epa.gov/otaq/models/biodsl.htm

[27] http://www.fact-index.com/b/bi/biodiesel.html#History

[28] http://www.fact-index.com/m/ma/main_page.html

[29] http://www.fao.org

[30] http://www.liquid-biofuels.com/pub.htm

[31] http://www.soypower.net/BiodieselPDF/BiodieselEmissions.pdf

[32] Intosh, C. S., Withers, R. V., Smith, S. M., 1982. The Economics of On-Farm Production and Use of Vegetable Oils for Fuel, Paper No.177, *Proceedings of the International Conference on Plant and Vegetable Oils as Fuels,* Holiday Inn Fargo North Dakota.

[33] Kaltschmitt, M., Reinhardt, G.A., 1997. *Nachwachsende Energieträger - Grundlagen,* Verfahren, ökologische Bilanzierung. Vieweg-Verlag. Braunschweig.

[34] Keys, J.D. 1976, *Chinese herbs, their botany, chemistry, and pharmacodynamics,* Chas. E. Tuttle Co., Tokyo.

[35] Krawczyk, T., "Biodiesel – Alternative fuel makes inroads but hurdles remain," *INFORM* vol. 7, no. 8(1996), pp.801-815, American Oil Chemist Society.

[36] Kusdiana D, Saka S. Effects of water on biodiesel fuel production by supercritical methanol treatment. *Biores Technol* 2004; 91: 289-295.

[37] Kusdiana D, Saka S. Kinetics of transesterification in rapeseed oil to biodiesel fuels as treated in supercritical methanol. *Fuel* 2001; 80: 693–698.

[38] Lin, C.Y., Lin, H.A, 2005, Diesel engine performance and emission characteristics of biodiesel produced by the peroxidation process, Fuel xx (2005) 1-8.

[39] Lyell, K., 2003. Design of a Methanol Extraction Process for Bio-Diesel Production, Fall 2003, Austin Bio-Fuels, LLC, Texas.

[40] Ma F, Hanna MA. Biodiesel production: a review. Biores Technol 1999, 70:1-15.

[41] Ma, F., "Biodiesel fuel: The transesterification of beef tallow." PhD dissertation, *Biological Systems Engineering*, University of Nebraska-Lincoln (1998).

[42] Ma, F., Clements, L.D., Hanna, M.A, 1998. Biodiesel fuel form animal fat. Ancillary studies on transesterification of beef tallow, *Ind. Eng. Chem. Res.* vol.37 (1998b), pp. 3768-3771.

[43] Ma, F., Hanna, M.A., Biodiesel production: a review, *Biosource Technology* vol.7 (1999), pp.1-15.

[44] Noureddini H, Gao X, Philkana RS. Immobilized Pseudomonas cepacia lipase for biodiesel fuel production from soybean oil, *Biores Technol* 2005; 96:769–777.

[45] O'Brien, R., 1998, Fats and Oils, *Formulating and Processing for Applications*, 667 p., Technomic Publishing AG, Basel, Switzerland

[46] Oda M, Kaieda M, Hama S, Yamaji H, Kondo A, Izumoto E, Fukuda H. Facilitatory effect of immobilized lipase-producing *Rhizopus oryzae* cells on acyl migration in biodiesel-fuel production. *Biochem Eng J* 2004; 23:45–51.

[47] Ortiz-Canavate, J., Hernanz, J.L., 1999. Energy for Biological Systems, Energy Analysis and Saving, *CIGR Handbook of Agricultural Engineering, Energy & Biomass Engineering,* pp.13-42, Published by ASAE, USA.

[48] Ozturk, O, 2004. *Aspir Tarımının Önemi ve Orta Anadolu Şartlarında Yetiştirme İmkanları, Konya Ticaret Borsası,* April 2004, Year: 7, N. 17, pp. 54-60 (Turkish), Konya, Turkey.

[49] Pischinger GM, Falcon AM, Siekmann RW, Fernandes FR. Methylesters of plant oils as diesels fuels, either straight or in blends. Vegetable Oil Fuels, ASAE Publication 4-82, *Amer Soc Agric Engrs St. Joseph,* MI, USA, 1982.

[50] Pryor, R. W., M. A. Hanna, J. L. Schinstock, and L. L. Bashford. 1982. Soybean oil fuel in a small diesel engine. *Trans ASAE* 26:333-338.

[51] Raghu, J.S. and Sharma, S.R. 1978, Response to irrigation and fertility levels of *safflower. Indian J. Agron.* 23(2):93-97.

[52] Riva, G., Sissot, F., 1999. Vegetable Oils and Their Esters (biodiesel), CIGR *Handbook of Agricultural Engineering, Energy & Biomass Engineering,* pp. 164-201, ASAE, USA.

[53] Schwab AW, Bagby MO, Freedman B. Preparation and properties of diesel fuels from vegetable oils. *Fuel* 1987;66:1372–1378.

[54] Shieh C-J, Liao H-F, Lee C-C. Optimization of lipase-catalyzed biodiesel by response surface methodology. *Bioresource Technology* 2003;88:103–106.

[55] Smith, J.R., 1985. "Safflower: Due for a Rebound", *J. Am. Oil Chem. Soc.,* 62(9):1286-1291.

[56] Sonntag, N.O.V., 1979. *"Composition and Characteristics of Individual Fats and Oils" in Bailey's Industrial Oil and Fat Products,* Vol. 1, 4th Edition, D. Swern, ed. New York, NY: A Wiley- Interscience Publication, pp. 398-403.

[57] Srivastava A, Prasad R. Triglycerides-based diesel fuels. *Renew Sustain Energy Rev* 2000;4:111–133

[58] Tickell, J., 2000.*From the Fryer to the Fuel Tank the Complete Guide to Using Vegetable Oil as an Alternative Fuel.* Tickell Energy Consulting, USA

[59] Tyson, K.S., 2004, *Biodiesel Handling and Use Guidelines,* DOE/GO 102004-1999 Revised November 2004; U.S. Department of Energy.

[60] Weiss, T.J., 1983. "Commercial Oil Sources", in *Food Oil and Their Uses,* Second Edition.

[61] Zhang Y, Dub MA, McLean DD, Kates M. Biodiesel production from waste cooking oil: 2. Economic assessment and sensitivity analysis. *Biores Technol* 2003; 90:229–240.

In: New Research on Biofuels   ISBN 978-1-60456-828-8
Editors: J. H. Wright and D. A. Evans   © 2008 Nova Science Publishers, Inc.

Expert Commentary

POLLUTION REDUCTION USING BIOFUELS: FROM THE LABORATORY TO THE REAL WORLD

C. Mazzoleni[1], H. Kuhns[2] and H. Moosmüller[2]

[1]Michigan Technological University, Houghton, MI, USA
[2]Desert Research Institute, Nevada System of Higher Education, Reno, NV, USA

Large scale production and consumption of biofuel is being promoted by many nations because of its potential for economic, political, and environmental benefits. Biofuels are partially renewable and substituting them for petroleum fuels may reduce dependency on imported oil and emissions of greenhouse gases and other pollutants. Of special interest is the possibility of widespread use of biofuels such as biodiesel in the transportation sector, especially if integrated with existing and new hybrid and plug-in technologies and emission control technologies such as particle traps.

Until recently, vehicle emission inventories have been mostly based on laboratory measurements, which have indicated that the use of biofuels can reduce the total emission of some pollutants dangerous to human health and the environment, such as particulate matter, hydrocarbons, and carbon monoxide. For example, an EPA comprehensive report averaging the results of many previous studies reports emission reductions of 10.1% in particulate matter, 21.1% in hydrocarbons and 11.0% in carbon monoxide, and an increase of 2.0% in nitrogen oxides for a blend of 20% biodiesel and 80% petroleum diesel. This study also reports that emissions generally decrease with increasing biodiesel fraction [Environmental Protection Agency 2002]. The vast majority of biofuel emission studies are performed on engine or chassis dynamometers under controlled laboratory conditions to carefully control engine-operating mode and to reduce confounding factors such as fuel quality, temperature, and humidity variability. Prior to emission testing, fuels are generally subjected to quality analysis. When fuels are not in compliance with the appropriate specifications, the low quality fuel is often discarded and the emissions test is not performed, skewing the databases toward the results obtained with high quality fuels.

Laboratory studies have been, and still are today, fundamental to elucidate emission factors for combustion processes under well-controlled conditions. However, the use of these laboratory-derived emission factors to accurately predict pollutant concentrations in highly impacted urban areas has often failed [National Research Council (U.S.). Committee on Vehicle Emission Inspection and Maintenance Programs 2000]. In addition, laboratory studies have frequently overestimated the beneficial effects of inspection and maintenance programs on urban air quality [National Research Council (U.S.), Committee on Vehicle Emission Inspection and Maintenance Programs 2001]. For these reasons there is a strong need to verify laboratory results under real world conditions and to study causes for these discrepancies and propose effective mitigation strategies. To represent real world conditions, emission measurements should at least explore the effects of the following variables: a) weather conditions (local meteorology, temperature, humidity, etc.); b) driving behavior and engine load; c) road condition; d) vehicle condition (age, tampering, and maintenance); e) non-normality of emission distribution across large fleets; f) fuel quality, composition, and additives. A realistic assessment of each of these many variables is difficult and expensive to represent in detailed laboratory studies. These data can be obtained in the real world with available technology which allows characterizing emissions for hundreds of vehicles in a cost effective way and in short time. Remote sensing and other real world measurement techniques, such as tunnel, on-board, road-side, and chase studies have been successfully used to evaluate emission reduction strategies and the accuracy of mobile emissions models [Bishop and Stedman, 2008, Bishop et al.1989, Bishop and Stedman 1996, Frey et al. 2003, Gertler and Pierson 1996, Jimenez 1999, Kuhns et al. 2004, Mazzoleni et al. 2004, Mazzoleni et al. 2004b, Pierson and Brachazek 1983, Sagebiel et al. 1997, Stedman 1989, Walsh at al. 1996, Herndon et al. 2005]. It is important to apply these techniques to evaluate the effectiveness of biofuel use in reducing pollutants in the real world.

A recent remote sensing case study [Mazzoleni et al., 2007] provided evidence of emission increases when substituting petroleum diesel with a biodiesel blend. Specifically, particulate matter emissions from school buses (see figure 1) significantly increased (up to a factor of 1.8) after the switch from 100% petroleum diesel to a B20 blend (20% biodiesel and 80% petroleum diesel). Cold-start CO emissions and hot-stabilized HC emissions were also higher with B20 while other tailpipe emissions were not significantly different. In this specific case, the introduction of a biodiesel blend increased overall particulate matter emissions and may have been detrimental to the local air quality. This study shows that the emission increases were likely due to the varying and somewhat low quality commercial biodiesel fuel sold in the region during the field campaign. In particular, high level of glycerin was found in the fuel (free glycerin in excess of 0.05% by weight). It has previously been shown that excess glycerin in the fuel may cause fuel system failure and poor combustion characteristics [NREL 2006]. Glycerin is a major byproduct of the trans-esterification process used to make biodiesel. Elevated free glycerin levels are caused by failure to completely separate the glycerin byproduct from the biodiesel product during production. Biodiesel production is a simple process which can be performed on small scale, even at home by individuals with little technical knowledge and lacking quality control. For example, a simple Internet search for recipes on how to produce biodiesel at home yields a large number of results. Unfortunately, such production methods may not properly and completely separate the glycerin, potentially resulting in higher emissions of particulate matter.

Figure 1. School bus powered with B20 (20% biodiesel and 80% petroleum diesel). The school bus was passing through an experimental set-up for the measurement of on-road emissions. Meridian, Idaho, March 2004 [Mazzoleni et al., 2007].

During a recent nationwide survey of 27 biodiesel and 50 B20 blends from blenders and distributors in the USA it was found that 85% of the biodiesel samples met the ASTM (American Society for Testing and Materials) D6751 biodiesel standard requirements even if only one sample met the European oxidation stability requirement. The remaining 15% samples were non compliant with USA standard for at least one regulated fuel property, this is a small but non-negligible fraction of the biodiesel sold in USA [McCormick et al. 2005].

Diesel particulate matter is not only dangerous to human health, but also contains a large fraction of light absorbing aerosol, often called black carbon. Light absorbing aerosol acts as strong absorber of solar radiation in the atmosphere, causing local heating. Light absorbing aerosols, in contrast to CO_2, are short lived and therefore are inhomogeneously distributed in the atmosphere. Light absorbing aerosols can produce local atmospheric heating while causing surface cooling and therefore changing the atmospheric stability. These particles can potentially contribute to regional and global climate changes including atmospheric warming and precipitation inhibition. Recent studies, for example, found strong relations between arctic radiative forcing, Himalayan glaciers melting, and light absorbing aerosol concentrations [McConnell et al. 2007, Ramanathan et al. 2007]. Additionally, model studies show how reducing black carbon could be an effective and rapid way to mitigate climate change [Jacobson 2002, Bond 2007]. Therefore, low quality biofuel usage could potentially exacerbate climate change through increased particulate matter emissions despite potentially lower greenhouse gas emissions.

To achieve and maximize the potentially significant benefits of biofuel usage, it is therefore imperative to ascertain biofuel quality and performance under real world conditions.

Further, it is important to design effective strategies for commercial fuel quality control and to design experiments to continuously monitoring real world effectiveness of these fuels in pollution and greenhouse gas emission reductions. The prefix "bio" is not sufficient alone, beyond rhetoric, economic interests, and politics, to guarantee the effectiveness of a biofuel in reducing human impact on environment and human health.

REFERENCES

Bishop, G. A.; Starkey, J. R.; Ihlenfeldt, A.; Williams, W. J. and Stedman, D. H. (1989). IR long-path photometry: A remote sensing tool for automobile emissions. *Anal. Chem.*, *61*, 671A-77A.

Bishop, G. A. and Stedman, D. H. (1996). Measuring the emissions of passing cars. *Acc. Chem. Res.*, *29*, 489-95.

Bishop, G. A. and Stedman, D. H. (2008). A decade of on-road emissions measurements. *Environ. Sci. Technol.* In press.

Bond, T. C. (2007). Can warming particles enter global climate discussions? *Environ. Res. Lett.*, *2*, 1-9.

Environmental Protection Agency, U. S. (2002). A comprehensive analysis of biodiesel impacts on exhaust. Available on-line: http://www.epa.gov/otaq/models/analysis/biodsl/p02001.pdf.

Frey, H. C.; Unal, A.; Rouphail, N. M. and Colyar, J. D. (2003). On-road measurements of vehicle tailpipe emissions using a portable instrument. *J. Air Waste Manag. Assoc*, *53*, 992-1002.

Gertler, A. W. and Pierson, W. R. (1996). Recent measurements of mobile source emission factors in North American tunnels. *Sci. Total Environ.*, *190*, 107-113.

Herndon, S. C.; Jayne, J. T.; Zahniser, M. S.; Worsnop, D. R.; Knighton, B.; Alwine, E.; Lamb, B. K.; Zavala, M.; Nelson, D. D.; McManus, J. B.; Shorter, J. H.; Canagranatna, M. R.; Onasch, T. B. and Kolb, C. E. (2005). Characterization of urban pollutant emission fluxes and ambient concentration distributions using a mobile laboratory with rapid response instrumentation. *Faraday Discuss.*, *130*, 327-39.

Jacobson, M. Z. (2002). Control of fossil-fuel particulate black carbon and organic matter, possibly the most effective method of slowing global warming. *J. Geophys. Res.*, *107*, ACH16-1-22.

Jimenez, J. L. (1999). *Understanding and quantifying motor vehicle emissions with vehicle specific power and tildas remote sensing, Ph.D. Thesis.* Boston, USA: Institute of Technology.

Kuhns, H. D.; Mazzoleni, C.; Moosmüller, H.; Nikolic, D.; Keislar, R. E.; Barber, P. W.; Li, Z.; Etyemezian, V. and Watson, J. G. (2004). Remote sensing of PM, NO, CO and HC emission factors for on-road gasoline and diesel engine vehicles in Las Vegas, NV. *Sci. Total Environ.*, *322*, 123-37.

Mazzoleni, C.; Kuhns, H. D.; Moosmüller, H.; Keislar, R. E.; Barber, P. W.; Robinson, N. F. and Watson, J. G. (2004). On-road vehicle particulate matter and gaseous emission distributions in Las Vegas, Nevada, compared with other areas. *J. Air Waste Manag. Assoc.*, *54*, 711-26.

Mazzoleni, C.; Moosmüller, H.; Kuhns, H. D.; Keislar, R. E.; Barber, P. W.; Nikolic, D.; Nussbaum, N. J.; Watson, J. G. (2004b). Correlation between automotive CO, HC, NO, and PM emission factors from on-road remote sensing: implications for inspection and maintenance programs. *Transp. Res. Part D, 9*, 477-496.

Mazzoleni, C.; Kuhns, H. D.; Moosmüller, H.; Witt, J.; Nussbaum, N. J.; Chang, M.-C. O.; Parthasarathy, G.; Nathagoundenpalayam, S. K. K.; Nikolic, G. and Watson, J. G. (2007). A case study of real-world tailpipe emissions for school buses using a 20% biodiesel blend. *Sci. Total. Environ., 385*, 146-159.

McConnell, J. R.; Edwards, R.; Kok, G. L.; Flanner, M. G.; Zender, C. S.; Saltzman, E. S.; Banta, J. R.; Pasteris, D. R.; Carter, M. M.; Kahl, J. D.W. (2007). 20th-century industrial black carbon emissions altered arctic climate forcing. *Science, 317*, 1381-1384.

McCormick, R. L.; Allemann, T. L.; Ractliff, M.; Moens, L. and Lawrence, R. (2005). Survey of the quality and stability of biodiesel and biodiesel blends in the United States in 2004. Technical Report NREL. Document available on-line at http://www.nrel.gov/docs/fy06osti/38836.pdf.

National Research Council (U.S.). (200). *Committee on vehicle emission inspection and maintenance programs. Modeling mobile-source emissions.* Washington, D.C, USA: National Academy Press.

National Research Council (U.S.). (2001). *Committee on vehicle emission inspection and maintenance programs. Evaluating vehicle emissions inspection and maintenance programs.* Washington, D.C., USA: National Academy Press.

NREL. (2006). Biodiesel handling and use guidelines. Document available on-line at http://www.nrel.gov/vehiclesandfuels/npbf/pdfs/40555.pdf.

Pierson, W. R. and Brachazek, W. W. (1983). Particulate matter associated with vehicles on the road. II. *Aerosol Sci. Technol., 2*, 1-40.

Ramanathan, V.; Ramana, M. V.; Roberts, G.; Kim, D.; Corrigan, C.; Chung, C.; Winker, D. (2007). Warming trends in Asia amplified by brown cloud solar absorption. *Nature, 448*, 575-8.

Sagebiel, J. C., Zielinska, B.; Walsh, P. A.; Chow, J. C.; Cadle, S. H.; Mulawa, P. A.; Knapp, K. T.; Zweidinger, R. B. and Snow. R. (1997). PM-10 exhaust samples collected during IM-240 dynamometer tests of in-service vehicles in Nevada. *Environ. Sci. Technol., 31*, 75-83.

Stedman, D. H. (1989). Automobile carbon monoxide emission. *Environ. Sci. Technol., 23*, 147-49.

Walsh, P. A.; Sagebiel, J. C.; Lawson, D. R.; Knapp, K. T. and Bishop, G. A. (1996). Comparison of auto emission measurement techniques. *Sci. Total. Environ., 190*, 175-80.

I

J

K

N

O